EAST ASIAN HISTORICAL MONOGRAPHS
General Editor: WANG GUNGWU

The Élite and the Economy in Siam
c.1890–1920

EAST ASIAN HISTORICAL MONOGRAPHS
General Editor: WANG GUNGWU

The East Asian Historical Monographs series has, since its inception in the late 1960s, earned a reputation for the publication of works of innovative historical scholarship. It encouraged a generation of scholars of Asian history to go beyond Western activities in Asia seen from Western points of view. Their books included a wider range of Asian viewpoints and also reflected a stronger awareness of economic and socio-cultural factors in Asia which lay behind political events.

During its second decade the series has broadened to reflect the interest among historians in studying and reassessing Chinese history, and now includes important works on China and Hong Kong.

It is the hope of the publishers that, as the series moves into its third decade, it will continue to meet the need and demand for historical writings on the region and that the fruits of the scholarship of new generations of historians will reach a wider reading public.

Other titles in this series are listed at the end of the book.

The Élite and the Economy in Siam
c.1890–1920

Ian Brown

SINGAPORE
OXFORD UNIVERSITY PRESS
OXFORD NEW YORK
1988

Oxford University Press
Oxford New York Toronto
Delhi Bombay Calcutta Madras Karachi
Petaling Jaya Singapore Hong Kong Tokyo
Nairobi Dar es Salaam Cape Town
Melbourne Auckland
and associated companies in
Berlin Ibadan

Oxford is a trade mark of Oxford University Press

© Oxford University Press Pte. Ltd. 1988

Published in the United States by
Oxford University Press, Inc., New York

All rights reserved. No part of this publication may be reproduced, stored in a retrieval system, or transmitted, in any form or by any means, electronic, mechanical, photocopying, recording or otherwise, without the prior permission of Oxford University Press

ISBN 0 19 588877 4

British Library Cataloguing in Publication Data

Brown, Ian, 1947–
The élite and the economy in Siam
c.1890–1920. — (East Asian historical monographs).
1. Thailand — Economic conditions
I. Title   II. Series
330.9593'035   HC445

ISBN 0-19-588877-4

Library of Congress Cataloging-in-Publication Data

Brown, Ian, 1947–
The élite and the economy in Siam, c. 1890–1920 / Ian Brown.
p. cm. — (East Asian historical monographs)
Bibliography: p.
Includes index.
ISBN 0-19-588877-4 (U.S.):
1. Thailand — Economic policy.
2. Elite (Social sciences) — Thailand — History — 19th century.
3. Elite (Social sciences) — Thailand — History — 20th century.
4. Thailand — Politics and government.   I. Title.   II. Series.
HC445.B76   1988   87-31338
338.9593 — dc19   CIP

Printed in Singapore by Kim Hup Lee Printing Co. Pte. Ltd.
Published by Oxford University Press Pte. Ltd.,
Unit 221, Ubi Avenue 4, Singapore 1440

For my parents

Acknowledgements

I wish to thank those individuals and institutions who have assisted me in the preparation of this study. Financial support for research visits to Thailand in 1978 and 1982 was provided by the Social Science Research Council (now the Economic and Social Research Council) and the School of Oriental and African Studies. The research itself was greatly facilitated by the unfailing helpfulness and encouragement of the staff of the National Archives in Bangkok, including, notably, Kanittha Wongpanit. The Institute of Southeast Asian Studies in Singapore gave me a valuable opportunity to complete my preliminary writing in late 1981.

Manas Chitakasem, one of my early teachers of Thai and a long-established friend, helped me unravel the nuances contained in the more impenetrable of my Thai materials. His patience was remarkable. An early draft of this study was read by Craig Reynolds and the late Neil Charlesworth. Their sharp criticisms, gently delivered, were invaluable—if this study has merit, it owes much to them. Neil's tragically early death in 1986 still leaves a painful sadness. Malcolm Falkus, Philip Stott, Kaoru Sugihara, and Shinya Sugiyama each offered valuable comment on particular passages and arguments. I remain responsible for all errors of fact and interpretation.

Over almost two decades Manoon and Yupha Chitakasem, and Penporn Satienswasdi have provided me with such hospitality and friendship as to make Bangkok, at its oppressive worst, bearable, and at its best, immensely enjoyable. For almost a decade my wife, Rajes, has prodded me to write more quickly. Over the last few years my sons, Andrew and Alasdair, have frequently asked why it was that even on the most glorious summer day this book appeared to hold a greater attraction for me than the prospect of cricket in the park. My explanation rarely convinced them; I now want them to know that I too was seldom convinced. And for much longer my parents have been constant in their encouragement of my stumbling academic ambitions. This book is dedicated to them.

Contents

Acknowledgements   vii
Notes   x

Introduction: The Élite and the Economy in Siam, c.1890–1920: The Problem   1
1. Irrigation in the Lower Čhaophrayā Plain   8
2. The State and the Rice Economy: Technical Change and Economic Depression   60
3. The Enclaves: Tin and Teak   94
4. The Élite and the Early Development of Indigenous Banking   125
5. The State and the Promotion of Economic Diversification   151
Conclusion: The Élite and the Economy in Siam, c.1890–1920   170

Select Bibliography   184
Index   192

Notes

Transcription

IN this study, the 'General System of Phonetic Transcription of Thai Characters into Roman' (as set out in the *Journal of the Thailand Research Society*, vol. 33, pt. 1 (March 1941), pp. 49–53) has been employed. There are two exceptions. First, the distinction between long and short vowel is indicated. Second, accepted usage has been followed in the case of some personal names (thus Chulalongkorn rather than Chulālongkǫn) and in the case of Thai words which have become familiar in English (thus baht rather than bāt).

Chronology

From the late 1880s to 1911, the Siamese chronological system was the *Rattanakōsinsok* (r.s.), the Bangkok Era (r.s. + 1781 = AD). In 1911 it was replaced by the *Phutthasakkarāt* (p.s.), the Buddhist Era (p.s. − 543 = AD). In the period covered by this study, the official Siamese year ran from 1 April to 31 March. Thus to convert, for example, the Siamese year r.s. 121 to the Western calendar, it has been necessary to use the cumbersome form 1902/3; p.s. 2456 would be 1913/14.

Introduction:
The Élite and the Economy in Siam,
c.1890–1920: The Problem

THE kingdom of Siam occupies a rare position in nineteenth- and twentieth-century Asia and Africa, being among the very few states in those two continents which did not experience Western rule during the age of high imperialism. Within South-East Asia, Siam was unique in this respect. Thus looking to her western border, Siam witnessed the gradual collapse of the Burmese kingdom before British power during the course of the nineteenth century, a process initiated by the first Anglo-Burmese War in 1824–6 and completed by a third outbreak of hostilities from late 1885. To the east, the kingdom saw the first advance of French rule, into the eastern provinces of Cochin-China, in 1858–62, and then the expansion of French administration to all the Indo-China territories by 1907. Looking to the peninsular south, the newly-founded Bangkok administration had seen the acquisition of Penang by the British in 1786 and the subsequent creation of an informal European imperium over the adjoining Malay States—and then from 1874, Siam watched the gradual consolidation of British formal administration in the Malay States, a process which brought all the territory to the south of the kingdom under British rule by the outbreak of the European war in 1914. Siam herself did not entirely escape these European territorial encroachments. Most notably, in 1893 the kingdom was forced to abandon to the French all claims on territory to the east of the Mekong River, whilst in 1909 she was pressured into transferring the four Malay States of Kedah, Perlis, Kelantan, and Trengganu to British authority. But these territorial sacrifices affected only the peripheral areas of the kingdom. Although the formal political independence of Siam was frequently threatened by strong diplomatic and occasionally military action on the part of Britain and, in particular, France at the turn of the century, by the time of the outbreak of the Great War in Europe in 1914, which effectively brought to an end European territorial acquisitions in the non-European world, the heartland of the kingdom of Siam and the major part of its outlying regions remained firmly under the authority of the indigenous administration.

The maintenance of Siamese formal independence in the late nineteenth and early twentieth centuries inevitably had a major impact on the political and social evolution of the kingdom in that period, and continues to exert a powerful influence on the contemporary Thai polity and society. Perhaps most importantly, Siam's avoidance of Western rule ensured the survival of the monarchy, indeed its preservation in a more powerful form, at least until absolute kingly authority was overthrown in a civilian–military coup in June 1932. Thus, whilst the final British advance into Upper Burma in late 1885 had been swiftly followed by the extinction of the Konbaung dynasty, and the imposition of French rule in the Indo-China territories had been accompanied by the emasculation of the indigenous courts, the later decades of the nineteenth century witnessed a very pronounced strengthening of the political power and religious authority of the Chakri kings. This in turn had far-reaching political and social implications for the kingdom. In particular the strengthening of the traditional focus of secular and religious authority, combined with the simple fact that the population of Siam continued to be governed at all levels by an indigenous administration, was one important reason why in the colonial era in South-East Asia, the kingdom did not experience the major political and societal tensions which afflicted, for example, Burma and Vietnam. If there was no colonial authority, there could be no anti-colonial upheaval.[1]

But turning to examine the specifically economic changes experienced by Siam during the decades of late European imperialism, the first impression is not of a distinctive Siamese experience but of a striking similarity between the evolution of the economy of Siam and that of the economies of British Burma and French Indo-China in the period beginning from the middle of the nineteenth century. In the years between 1850 and 1870 all three countries began to experience, in the delta areas of the Irrawaddy, the Čhaophrayā, and the Mekong, an expansion of rice cultivation for export that was, within three decades, to make mainland South-East Asia pre-eminent in the world rice trade. In each of the three deltas the actual cultivation of rice was undertaken almost exclusively by the indigenous population, whilst the financing, transportation, milling, and export of the crop increasingly became the preserve of the immigrant Asian communities which grew rapidly in size and economic influence throughout mainland South-East Asia in the second half of the nineteenth century—essentially Indians in Burma, and Chinese in Siam and Indo-China. The expansion of rice exports was accompanied by a growth in production for export of other primary commodities, including teak, tin, and rubber, although in these sectors non-indigenous enterprise—either the immigrant Asian communities or Europeans—was dominant. As export earnings for British Burma, Siam, and French Indo-China rose rapidly from the middle of the nineteenth century, they permitted, and were in turn stimulated by, an equally dramatic increase in imports from the industrialized world. The imports into each of these three states included considerable shipments of capital goods for certain of the local export sectors, but this trade was dominated by cotton piece goods whose appearance in local markets was almost certainly an important incentive for indigenous cultivators to expand production for export. Furthermore, as the indigenous populations of British Burma, Siam, and French Indo-China devoted an increasing proportion of their resources to production for export, and as imported consumer items entered the domestic market in growing quantities, indigenous handicraft and cottage industry in each of these countries declined, possibly in absolute terms but certainly in relation to total local consumption.

Marked similarities could also be seen in the economic life of the rapidly growing primate cities of later nineteenth-century mainland South-East Asia. In an economic sense there was little to distinguish Bangkok from either Rangoon or Saigon. In each there were the major concentrations of Chinese or Indian immigrants, engaged primarily, whether as labourers, small-scale traders, or influential merchants, in the processing and shipment of the country's primary exports and in the receipt and distribution of imports. In addition, in each of these three cities could be found major Western commercial institutions—European exchange banks, shipping offices, and merchant houses. Moreover, by the later nineteenth century, Rangoon, Bangkok, and Saigon had each become the focus of an intricate internal communications network—involving railways but primarily waterways—that brought considerable economic coherence to the immediate hinterland of the port.

Finally, it is possible to note strong similarities in a number of important areas of government economic administration and legislation between British Burma, Siam, and French Indo-China in this period. Two brief examples may suffice. In all three countries, administrative and legislative action was undertaken to provide more clearly defined property rights in agricultural land, primarily in order to facilitate the expansion of rice cultivation. Second, by the early twentieth century, the European administrations in Burma and Indo-China, and the indigenous administration in Siam, had each taken broadly similar legislative and administrative measures to regulate the extraction of important local wasting resources, including, notably, teak.

These similarities—in broad outline and often in detail—between the economic structures, institutions, administration, and life of each of the countries of mainland South-East Asia should cause little surprise. In essence they were simply a reflection of the increasingly close integration of the region into the international economy from the middle of the nineteenth century. The fact that in Burma and Indo-China that integration was effected by annexation and the forced imposition of European rule, whilst in Siam it was achieved under the authority of the indigenous élite, appears, at least at this initial level of analysis, to have been of little consequence.

Contemporary Western observers wrote approvingly of the economic changes which had occurred in mainland South-East Asia in the second half of the nineteenth century,[2] confidently seeing them as part of a world-wide advance in human industry and commercial intercourse which would raise man's material condition. Within a span of just five decades, vast tracts of intensely inhospitable jungle had been cleared and turned to the cultivation of rice. Internal communications, which in 1850 had been slow, unpredictable, and occasionally even hazardous, had, by the turn of the century, improved almost beyond recognition with the cutting of numerous new canals and the construction of railways. The ports of Rangoon, Bangkok, and Saigon which in 1850 carried little international trade were by 1900 major centres of commerce in the East. The banks of the Irrawaddy, Čhaophrayā, and Mekong Rivers were, near their mouth, lined with steam-powered rice-mills and sawmills, whilst their lower reaches carried countless rice barges and teak rafts. At their wharves lay numerous local craft and ocean-going vessels, loading with primary exports or discharging their cargoes of foreign cotton shirtings, machinery, and immigrants from India or China, whilst in the main commercial districts of Bangkok, Rangoon and Saigon, traders, bankers and merchants maintained a relentless activity. Yet, if these economic changes excited the imagination and even pride of the early twentieth-century observer, they have almost invariably provoked a more sombre assessment from those of the present day. In the dramatic expansion of rice exports from mainland South-East Asia from the second half of the nineteenth century are now seen the roots of a dangerous dependence on primary commodity production and trade, whilst in the equally dramatic expansion of imports during the same period is seen the destruction of local cottage industries and handicrafts. Or again, the rapid growth of Western merchant firms, banks, extractive companies, and shipping lines throughout mainland South-East Asia from the middle of the century is now viewed not as proof of economic progress, but as manifestation of a deleterious domination of the local economies by foreign capital. In general terms, the economic structures which emerged in mainland South-East Asia during this period—indeed which emerged in the broadest sense throughout virtually the whole non-European world during the decades of high imperialism—are now widely regarded as having condemned those areas to long-term economic dependency and stagnation.[3]

There are, therefore, two preliminary observations which may be made. The first is that despite the radically different political circumstances of Siam from those of Burma and Indo-China in the later nineteenth and early twentieth centuries, the pattern of economic change which emerged in Siam was markedly similar to that which emerged in the two neighbouring colonial states. And second, that that pattern of economic change—in Siam as much as in Burma and Indo-China—was to the long-term economic disadvantage of the indigenous populations. It is from these two preliminary observations that the central issues to be considered in this book are derived. The most important of them may be stated as follows: to what extent did Siam, under the authority of an indigenous administration, have an opportunity to pursue an alternative pattern of economic change—once again in broad outline or in detail—in the late nineteenth and early twentieth centuries? From this, two further questions are drawn. To what extent did the indigenous Siamese administration perceive either the necessity for or the desirability of promoting a pattern of economic change in any way significantly different from the one that was emerging in the kingdom from the middle of the nineteenth century? To what extent was the Siamese administration of this period constrained by either external or internal circumstances (including the self-interest of its own administrative élite) from promoting an alternative pattern of economic change?

In summary, this study is concerned primarily with the Siamese administration's perception of and response to the economic changes which were occurring within the kingdom in the late nineteenth and early twentieth centuries, and with the constraints which shaped that perception and which moulded that response. Had the Siamese administrative élite of the early twentieth century, after five decades of increasing familiarity with the Western world, assumed the values, attitudes, and intellectual horizons of the West to the extent that, in the promotion of economic change, it pursued essentially the same measures and sought substantially the same objectives as a European colonial administration? To the extent that the Siamese administrative élite of this period pursued economic policies similar to those of a European colonial regime, did it do so by conviction or because it was in some way forced to adopt such measures? Alternatively, was the Siamese administrative élite of the later nineteenth and early twentieth centuries in any way critical of the pattern of economic change evolving in the kingdom; and if so, to what extent and in what manner was it constrained from promoting an alternative pattern of economic change?

In seeking to probe this broadside of related and overlapping questions, five specific areas of government economic policy and administration in the late nineteenth and early twentieth centuries will be examined. These concern proposals for the construction of large-scale irrigation facilities in the lower Čhaophrayā delta; the promotion of technical change in agriculture and the response of the administration to the rice economy in crisis; the exploitation of the kingdom's teak and tin resources; the early development of indigenous banking in the kingdom; and the promotion of economic diversification within Siam. It is recognized that a number of important areas of government economic policy in this period are omitted from the discussion or receive only cursory treatment. These include the government's policy towards the construction of railways,[4] its land administration,[5] and Siamese financial administration.[6] However, the principal concern of this study is not to provide a comprehensive account of Siamese economic and financial policy in this period but rather, as explained, to examine the perceptions and responses of the Siamese administrative élite of the late nineteenth and early twentieth centuries to the pattern and pace of the economic changes which were evolving in the kingdom.

For these purposes, the five areas of policy and administration noted above should provide sufficient insight.

The study will cover the period c.1890–1920. There are two reasons for this choice of period, one practical and one intellectual. The intellectual justification is that it was during the late nineteenth and early twentieth centuries that the pace of economic change within the kingdom was particularly rapid, and that the imperial threat to Siam was at its most intense. Understandably, it was thus also the period when the responses to and the reflections on the part of the Siamese administrative élite with respect to the intrusion of the West were particularly vigorous. The practical reason for the choice of this period is related to this last point, for it is only for these years that the Siamese administrative records exist, and are open to scholars, in sufficient quantities to make possible the detailed examination of élite perceptions and responses which is the central objective of this study.

---

1. For a notably perceptive analysis of the nature of political and societal change in Siam under the later Chakri kings see, Benedict R. O'G. Anderson, 'Studies of the Thai State: The State of Thai Studies', in Eliezer B. Ayal (ed.), *The Study of Thailand: Analyses of Knowledge, Approaches, and Prospects in Anthropology, Art History, Economics, History, and Political Science*, Athens, Ohio, 1978, pp. 193–247.

2. With regard to Siam see, for example, Robert Gordon, 'The Economic Development of Siam', *Journal of the Society of Arts*, vol. 39 (1891), particularly pp. 291–8; A. Cecil Carter (ed.), *The Kingdom of Siam*, New York and London, 1904; J. G. D. Campbell, *Siam in the Twentieth Century: Being the Experiences and Impressions of a British Official*, London, 1902, pp. 18–58 *passim*.

3. The literature here regarding specifically Siam is very extensive—and that with respect to the non-European world as a whole is simply immense. References are held over to the concluding chapter.

4. See, David F. Holm, 'The Role of the State Railways in Thai History, 1892–1932', Ph.D. diss., Yale University, 1977.

5. See, David Feeny, *The Political Economy of Productivity: Thai Agricultural Development 1880–1975*, Vancouver and London, 1982, pp. 93–8.

6. This area has been considered by the author at length elsewhere. Ian Brown, 'The Ministry of Finance and the Early Development of Modern Financial Administration in Siam, 1885–1910', Ph.D. diss., University of London, 1975. Parts of this discussion have appeared in, Ian Brown, 'British Financial Advisers in Siam in the Reign of King Chulalongkorn', *Modern Asian Studies*, vol. 12, pt. 2 (April 1978), pp. 193–215; Ian Brown, 'Siam and the Gold Standard, 1902–1908', *Journal of Southeast Asian Studies*, vol. 10, no. 2 (September 1979), pp. 381–99.

1
Irrigation in the Lower Čhaophrayā Plain

THERE are strong reasons for beginning this study with an examination of the response of the Siamese administration to a series of proposals, placed before it in the opening two decades of the twentieth century, for the construction of large-scale irrigation works in the lower Čhaophrayā plain. The first is that these works, if undertaken in full, would have had a very major impact on agriculture in Central Siam, both in raising the productivity of the land and in restructuring the pattern of cultivation and husbandry. The second reason, related in part to the first, is that a number of studies have already drawn from the irrigation policy of the Siamese administration in the early twentieth century, important insights into the government's economic perceptions and priorities in that period.

I

The construction of canals represented an important and long-established activity for the pre-modern Siamese state. According to one authority, between the reigns of Rāmāthibọ̄dī (1351–69), the founder of Ayudhya, and Rama III (1824–51), more than twenty large-scale canals were constructed in the kingdom, the majority on the orders of the King as state public works.[1] During the Ayudhya and early Bangkok periods, the principal objective of canal construction by the state was to improve inland water navigation, primarily in order to facilitate the movement of military troops and their supplies during periods of war, but also to ease the movement of corvée deliveries and foreign commerce. These large-scale canals were usually excavated by corvée labour, working under the authority of a high-ranking official. Although during the Ayudhya period, a large number of small water courses was also cut in the upper part of the delta in order to facilitate irrigation, they were cut on the initiative of local communities themselves. The pre-modern Siamese state placed relatively little emphasis in its large-scale hydraulic works on the maintenance or advancement of agricultural production.[2]

From the third reign of the Bangkok period, important changes began to appear in both the objectives of the Siamese state in constructing canals, and in the way in which the resources for excavation were organized. For example, in the excavation work itself, use was now made of wage labour drawn from the immigrant Chinese community rather than corvée labour drawn from the indigenous population, for the former was found to provide a notably more reliable and productive work-force. With respect to the objectives of canal construction, from the reign of Mongkut (1851–68), the excavation of major canals was undertaken to meet economic rather than military needs. Thus, in 1857 work was begun on the Mahāsawat canal which, running westward from Bangkok, was intended to facilitate the transportation of sugar-cane, rice, and other agricultural produce from the areas of Nakhǫn Chaisī and Nakhǫn Pathom to the capital. Similarly the Phāsī Charoen canal, which ran south-west from the capital to the Thā Chīn and which was completed in 1872, was designed primarily to facilitate the movement of sugar-cane into Bangkok. Both these canals also served to open up previously unclaimed lands for the cultivation of rice. In the decades which followed, as Siam strove to meet a strongly rising international demand for rice, this was to become a dominant objective in the cutting of new, major, canals.

The reign of Mongkut also saw the emergence of a problem which was to be arguably the most important influence in shaping the government's canal construction policy into the twentieth century—difficulties over finance. In the mid-nineteenth century the immediate problem was that the high costs involved in employing Chinese wage labour to excavate new canals could not be met in full by a government whose traditional financial machinery had recently been seriously undermined by the provisions of the Bowring Treaty signed in 1855.[3] Consequently, in the 1850s and 1860s it was necessary to seek capital for canal construction from other sources—from wealthy Chinese, noble officials, and from the King himself. The construction of the Mahāsawat and Phāsī Charoen canals will illustrate the point. In the case of the former, the greater part of the wages of the Chinese labourers engaged for the work was provided by the King from the confiscated property of an errant official of the previous reign. In the case of the Phāsī Charoen canal, construction was originally proposed by a Chinese opium and sugar-cane tax-farmer, Phra Phāsī Sombatbǫribūn (Pho Jim), who, recognizing that the government would be unable to meet the full costs of the project, suggested that excavation be financed by the imposition of a toll on boats using the canal, and by the establishment of gambling houses at Nakhǫn Chaisī and Samut Sākhǫn for limited periods.[4] Finally, it might also be noted that the excavation of the Damnoen Saduak canal, which linked the Thā Čhīn and Maeklǫng Rivers to the south-west of Bangkok and which was completed in 1868, was financed by Somdet Čhao Phrayā Sīsuriyawong (Chuang Bunnag), the Minister of Military Affairs, using funds appropriated from the balance of a sugar tax revenue which had been placed in the Minister's charge for the construction of a royal palace at Phetburī. Where the king, noble officials, or wealthy Chinese invested heavily in the construction of a canal, they invariably claimed ownership of the vast tracts of wasteland which lay adjacent to it. Thus it was that lands along the central part of the Mahāsawat canal were subsequently granted by Mongkut to his sons and daughters to hold as ricelands exempt from land tax, whilst Somdet Čhao Phrayā Sīsuriyawong distributed lands along the Damnoen Saduak canal to his wives, other relatives, and dependants, or sold plots directly for cultivation.

An important disadvantage of this method of financing canal excavation was that the major tracts of land claimed by the promoters were left largely uncultivated. In essence few members of the royal family, noble officials, or wealthy Chinese appear to have been able to command sufficient manpower (corvée peasants and debt slaves) to undertake rice cultivation on the whole of their newly-acquired lands. Consequently, the 1870s saw a further shift in government canal policy, which now sought the related objectives of curbing the acquisition of large landholdings adjacent to new canals by members of the royal family and noble officials, and of exacting from the peasants who were actually to cultivate newly-opened lands, part of the costs of canal excavation. Most prominently, the Regulation of Canal Excavation of 1877 laid down that when, in future, plans for the construction of a canal through an unsettled district were announced, interested parties could apply to government officials for specific land allocations along the proposed excavation. The cost of construction would then be estimated, and apportioned among the land applicants, who could meet their obligations either in the form of cash or in the form of labour in the actual excavation of the canal. To encourage further peasant interest in land reclamation, the 1877 Regulation also eased the conditions upon which land-deeds were fully secured by the cultivators, and granted a limited exemption from land tax with respect to the lands being settled through new canal excavations.

The policy of encouraging the cultivators themselves to reclaim lands adjacent to newly-constructed canals whilst requiring them to bear part of the costs of excavation did not survive long into the 1880s, for that decade saw a very marked reduction in government canal construction. Johnston suggests that this may have reflected in part the internal political tensions of the period, in that they may have impaired the ability of the administration to undertake major construction projects efficiently, or it may have reflected an awareness on the part of the government that the very substantial canal excavations of earlier decades had, for the time, opened up a larger area of land than the existing population of the Central Plain could cultivate.[5] In addition, as the canal policy of the 1870s still required the government to meet a major part of excavation costs, continued pressure on the administration's limited financial resources may have been a further important influence in reducing the government's initiatives in this area. But whatever the influences here, the 1880s saw a further major change in official canal policy. The government now sought to grant concessions to private individuals or companies (principally members of the royal family, senior officials, and influential Chinese), to excavate canals, under which the concessionaire was recognized as the owner of lands contiguous to the canal under construction. By the subsequent sale of those lands to cultivators, the concessionaire could expect to recover his excavation costs; and indeed secure a substantial profit.

Of the companies established in the 1880s and 1890s to take up canal excavation and land development concessions, by far the most prominent was the Siam Land, Canals, and Irrigation Company.[6] This company was founded by Phra-ongčhao Sai Sanitwongse, son of an important younger half-brother of Mongkut and Chulalongkorn's personal physician, in partnership with Joachim Grassi, an Italian architect and owner of a construction firm in Bangkok, and two other Siamese. Their contract with the government, signed in January 1889, gave the company a virtual monopoly in canal and land development projects throughout Siam for a period of twenty-five years. The company originally envisaged the excavation of a network of canals to cover the whole of the lower delta on both sides of the Čhaophrayā. The first segment to be undertaken was the Rangsit system, in an area immediately to the north-east of the capital between the Čhaophrayā and Bāngpakong Rivers. Excavation of the main Rangsit canal was approved by the government in 1890; serious construction began two years later, and was completed in 1896. By 1900, some 500,000 *rai* (1 *rai* = 1600 square metres) had been cleared for cultivation and when, a few years later, excavations were essentially completed, it was estimated that between 1,250,000 and 1,500,000 *rai* had been opened up. The Rangsit project was also notable for one important constructional innovation. Where the main canal joined the Čhaophrayā and Bāngpakong Rivers, lock gates were constructed, thus making it possible for the Rangsit system to retain the water which flowed into it. This facility greatly enhanced the attractiveness of the area to cultivators, such that in the 1890s and early 1900s the company fell far short of meeting all applications for land, despite the very considerable expansion in the cultivable area during that period.

However, despite the undoubted success of the Rangsit concession in terms of the opening up of new lands for the cultivation of rice, by the late 1890s, the administration had developed sufficient doubts with respect to contracting out canal excavation to prompt a further major change in policy. In part, the government came to recognize that in granting major canal excavation and land development concessions to private individuals and companies, it was permitting those interests to determine, in large part, the pattern of agricultural settlement in the delta. This major responsibility should rightly rest with the administration itself. Concern was also expressed during this period that the work of all the land development concessionaires, including the work of the Siam Land, Canals, and Irrigation Company, was continuously behind schedule, being delayed by serious disputes over land ownership and, in some cases, by lack of sufficient capital resources. In addition, it was felt in some quarters that the system of land development by private concession had allowed the companies to profit at the expense of both the administration and the cultivator. But most importantly, it was argued from the late 1890s that the contracting out of canal excavation had failed to provide a coordinated irrigation system in the lower delta. For example, although the Rangsit canal system could retain the water which flowed into it, there were no constructions in the lower delta which would first ensure an adequate supply of water for that district in all years. Moreover, the districts to the south which were dependent on water flowing through the Rangsit area could be seriously disadvantaged by the retention of water within that canal system. It was therefore clear that a marked degree of government initiative and intervention was necessary to secure an essential coordination of irrigation requirements in the lower delta. Finally, by the late 1890s, the administration had become concerned that the Rangsit canals were already showing signs of serious deterioration. It had become apparent that they had been cut too narrow in relation to their depth, and consequently had accumulated considerable deposits of silt within a few years of their completion. As the Siam Land, Canals, and Irrigation Company showed little inclination to redredge or repair these canals, particularly in those circumstances where the major part of the adjacent lands had already been sold, this was a further powerful argument for the re-establishment of state initiative and authority in the construction of the kingdom's canals.

II

During a visit to Europe in 1897, Chulalongkorn attempted to secure the services of a Dutch irrigation engineer.[7] However, nothing came of this initiative, partly because the Siamese Legation in Holland failed to pursue the matter, and partly because the Minister of Public Works in Bangkok apparently showed little interest in irrigation and consequently neglected the issue. This subject was raised again in late 1899, on this occasion by the Financial Adviser, Charles Rivett-Carnac. The Adviser noted that there was now a large accumulated balance in the Treasury, and he suggested that it could be used most prudently in promoting irrigation, and specifically in engaging a first-class irrigation engineer, preferably from India. Rivett-Carnac regarded irrigation canals as more important for the development of Siam than either railways or roads, and he declared that 'I should rejoice to see the day upon which His Majesty's Government resolutely set itself to constitute a Department of Irrigation'.[8]

Rivett-Carnac's enthusiasm for a state irrigation programme was shared by a number of prominent members of the Siamese administration. During the previous month, November 1899, the Minister of Agriculture, Čhaophrayā Thēwētwongwiwat, had rejected applications from four private companies and individuals, including the Siam Land, Canals, and Irrigation Company, for further concessions to cut canals.⁹ In referring these applications to the King, Čhaophrayā Thēwēt noted specifically the problems which had arisen in the Rangsit concession. The Minister's poor opinion of the work of the private canal concessionaires, and indeed his dissatisfaction with the state of the delta's canal system as a whole, was strengthened when he undertook inspection tours of the delta in April and October 1900.¹⁰ He reported that travel by boat in many parts of the Central Plain had become notably difficult because the canals were either silted up or had become clogged with weeds. Čhaophrayā Thēwēt's response to the disturbing failures of private canal excavation was to argue that the government should itself undertake construction, and, like Rivett-Carnac, he proposed that as a first step the administration should engage a specialist in irrigation engineering.¹¹ These views were shared by the King.¹² At the suggestion of the General Adviser, Rolin-Jacquemyns, it was decided to seek an irrigation engineer from the Netherlands East Indies.¹³ Negotiations involving the Dutch colonial administration began in early 1900 but it took over two years for the Siamese Government to reach agreement with a suitably qualified and experienced man, J. Homan van der Heide.¹⁴

In retrospect, it can be said that the contract which the Siamese administration signed with van der Heide in April 1902 defined his responsibilities far too loosely. As well as being required to establish an Irrigation Department and to assist in the implementation of any irrigation system adopted by the government, van der Heide was to 'prepare, under the instructions of the Minister [of Agriculture], technical reports about the best Irrigation systems for the Lower Provinces of Siam and estimates of their probable costs'.¹⁵ As later events were to show, van der Heide was a man who longed to work on an ambitious scale: asking him for his unfettered views on the 'best' irrigation systems for the lower delta without making reference to the financial resources of the government, was to invite misunderstanding and acrimony in the future.

Van der Heide arrived in Siam on 13 June 1902 and immediately embarked on a series of extensive inspection tours of the delta provinces that were to occupy him almost exclusively during his first few months in the kingdom.¹⁶ However, whilst he was engaged in this preliminary work, an event occurred in the north of the kingdom that was to have a crucial influence on the fate of van der Heide's irrigation proposals. On 25 July 1902, an armed force of Shans attacked the northern town of Phrāe, stormed government offices, and killed approximately twenty Siamese provincial officials.[17] Then, on 3 August, the Shans launched an abortive attack on Lampāng, to the west of Phrāe. The Bangkok Government immediately despatched troops from Phitsanulōk and Nakhon Sawan to quell the uprising. However, at that time construction of the northern railway had reached only Lopburī, and, consequently, the government troops were forced to march and boat up-stream to reach the disaffected areas. It was therefore several weeks before government forces re-entered Phrāe. The time taken to move a major force of soldiers into this critical border region, coupled with the suspiciously close interest which had been paid to the uprising by the French and British consuls at Nān, convinced the Bangkok authorities that it was essential to push ahead with construction of the northern railway as rapidly as possible, and instructions to this effect were given by Chulalongkorn in late August 1902. Yet, financing this major work proved difficult. Until that time the government's railway construction programme had been financed entirely from the administration's tax revenues, but the scale and urgency of the northern line project, coupled with rising demands for funds from other government departments, made a continuation of this policy impossible. In late October 1902, the Minister of Finance, Prince Mahit, was able to provide 300,000 baht to begin construction beyond Lopburī, but by the end of the year, Ministers in Bangkok were being forced to give serious consideration to the raising of a foreign loan.[18] In fact, there were very considerable misgivings within the Siamese administration over the possible political implications of the kingdom becoming a debtor of the European powers, and it was therefore not until November 1904 that the Council of Ministers, under pressure from an increasingly severe budget crisis, finally sanctioned the raising of a loan overseas. That loan, for £1 million, was floated jointly in London and Paris in March 1905. In summary, the latter half of 1902—the very months when van der Heide was conducting his preparatory investigations and writing his report—saw a major shift in the financial position of the Siamese Government. At the end of 1899 Rivett-Carnac had been able to point to a large accumulated balance in the Treasury, and suggest that it be used to promote irrigation. In January 1903, when van der Heide submitted his report, the Siamese administration had entered a long period of severe budgetary restraint during which there would be the most acute competition for scarce resources.

Van der Heide's *General Report on Irrigation and Drainage in the Lower Menam Valley* presented the most assured case for the construction of a large-scale irrigation system in Central Siam. In its opening pages, it established that in the lower Čhaophrayā plain, the rainfall in ordinary years was quite insufficient to meet the full requirements of rice cultivation, whilst in years of scanty rainfall, large-scale crop failure was almost inevitable. It was 'only in extremely rare cases, in very rainy years', that the rainfall was fully sufficient.[19] The existing canals in Lower Siam, even those within the Rangsit district, were incapable of making good this deficiency, for they were merely distribution channels. They could not increase the supply of water to the rice-fields of the delta in years of deficient rainfall.

Therefore the most powerful argument for the construction of a large-scale, effective irrigation system was that it would secure, in all years, a sufficient supply of water for rice cultivation in the Central Plain. Failure of the rice crop because of poor rains would be prevented. But this was by no means the only advantage. An irrigation system would prolong the time available for the cultivation of rice in each season by one and a half to two and a half months. This would not only enable each cultivator to take on substantially more land, but would also permit greater care to be taken in the cultivation of the crop, as well as allow the planting of those strains of rice which take longer to mature. Furthermore, an irrigation system would ensure that a substantial quantity of silt would be carried on to the rice-fields. It would enable villages to be protected from flooding, so making possible the planting of fruit trees, cultivation of garden produce, and the raising of cattle. By making water available in the dry season it would also permit the cultivation of a dry season crop such as maize, beans, cotton, or tobacco. In addition, an irrigation system would provide fresh drinking water throughout the year for the whole area covered by the scheme. And finally, by maintaining high water levels in the canals throughout the year, it would considerably improve communications in the Čhaophrayā plain.

As the basis of his scheme, van der Heide proposed the construction of a weir across the Čhaophrayā River at Chaināt, on the northern edge of the lower Central Plain. In addition, two main canals would be cut, running parallel with the Čhaophrayā River, and from these main channels, smaller canals would carry water to the different parts of the lower delta. Van der Heide estimated that the scheme would bring irrigation water to more than half the currently cultivated area of Siam. It would take approximately twelve years to complete construction, and the cost was provisionally estimated at 47 million baht.[20] Van der Heide maintained that this sum was comparatively modest. The costs of the project were low, he suggested, because the major part of the main canals and smaller distributaries would consist of existing water channels. In addition, the government could anticipate a considerable financial return from the project, indirectly by way of a general increase in revenues consequent upon the expansion of rice production, and directly, van der Heide proposed, by the imposition of irrigation charges and by the sale of newly-irrigated lands. These financial returns would be sufficient to cover not only the costs of maintaining and managing the irrigation works but a considerable part of the original outlay as well. However, van der Heide recognized that the government might still feel that it could not afford to construct the full scheme in the time proposed, and he therefore suggested that it would be possible to proceed with one at reduced capacity, at a cost of 28 million baht; at the same time he firmly held that it would be against the economic interests of the people and the long-term financial interests of the government to undertake anything but the full scheme. To underpin this point he argued that the value of lost crops in Central Siam in just one year of serious drought was likely to exceed the cost of the irrigation scheme at full capacity.

Van der Heide's report was an incisive analysis of Siam's irrigation requirements, and how they might be met; and the fact that the Dutch engineer was able to present his very substantial report within just six months of his arrival in the kingdom makes his achievement all the more notable. Nevertheless, his proposals were poorly received by the Siamese administration. The Minister of Foreign Affairs, Prince Devawongse, writing to Chaophrayā Thēwēt in May 1903, argued that although van der Heide's scheme would secure very substantial benefits for the kingdom, it would absorb such a large volume of government funds that insufficient resources would be left for other, more important projects.[21] It is clear that Prince Devawongse primarily had in mind the construction of railways, particularly the line to the north. Support for van der Heide might have been expected from Chaophrayā Thēwēt himself, partly because of his position as Minister of Agriculture but more particularly because of his earlier enthusiasm for a government irrigation scheme. However, that support was not forthcoming. In May 1903, Čhaophrayā Thēwēt removed from van der Heide's new contract, then under negotiation, a clause which indicated that the newly-established Irrigation Department would undertake the implementation of the Dutchman's report, and inserted the more general statement that the new department would be responsible for restoring and then maintaining the existing canals of the kingdom and, in a general sense, for the administration of the supply of irrigation water for cultivation.[22] Then in August 1903, the Minister informed the King that after a full consideration of van der Heide's report he believed that the most important immediate task of the government in this area would be to repair the existing canals whilst carrying out the preparatory work for the 'big scheme'.[23] Van der Heide had been asked to prepare estimates for these smaller projects for the year 1903/4. This he had done, but Čhaophrayā Thēwēt, by rejecting certain proposals, had been able to reduce the estimate further by almost a half. In addition, substantial reductions had been made in the expenditure estimates for irrigation for the four years from 1904/5.

It is clear that if even the Minister of Agriculture was not prepared to argue in support of van der Heide's proposals, then they would have little chance of being carried in the Council of Ministers. In mid-August 1903 a committee was established, comprising Čhaophrayā Thēwēt himself, Prince Devawongse, Prince Damrong (the Minister of the Interior), Prince Mahit (the Minister of Finance), and Prince Narit (the Minister of Public Works), to examine critically government expenditure on railways and irrigation.[24] Unfortunately, there appears to be no surviving record of the committee's deliberations. However, with respect to irrigation, approval was apparently given only to a series of modest measures as outlined earlier by Čhaophrayā Thēwēt. It is almost certain that a final decision on van der Heide's 'big scheme' was postponed.[25]

During the following three years van der Heide, as Director of the Canal Department, was forced to content himself with administering these minor works. Restoration work was carried out on a number of major canals, and then work was begun on constructing locks on the Sāen Sāep, Phāsī Čharoen, Damnoen Saduak, Prawēt, and Samrōng canals in order to improve both water communications in the lower Central Plain and the supply of water for rice cultivation.[26] These projects necessitated the purchase of six dredgers and the recruitment of a substantial European staff—as early as September 1904 there were eight European engineers and three European surveyors employed in the Canal Department.[27]

This work was, of course, of considerable value, and it could be argued that in improving the existing canals in the lower delta the Canal Department was carrying out essential preliminary work for the 'big scheme'. However, these limited projects were never likely to satisfy van der Heide's vision, and in early 1906, he submitted further proposals to the government.[28] They were not new. In essence van der Heide proposed the execution of certain of the specific projects contained in the 1903 report, the clear implication being that in time the government would complete the full scheme as originally presented. Specifically, he proposed the construction of the weir at Chaināt, completion of the east bank works at reduced capacity, and execution of a number of small works on the west bank. The programme would take four years to complete at an annual cost rising from almost 4 million baht in the first year, to 7 million baht in the last. The total cost was estimated at a little over 24 million baht.[29]

In 1908, van der Heide was to suggest that his 1906 proposals had been drawn up on instructions from Čhaophrayā Thēwēt, who had informed him that the King 'had decided to have the irrigation scheme and the water supply scheme for Bangkok jointly carried out without delay'.[30] In fact, it is extremely unlikely that Chulalongkorn had ever made such a commitment, for in 1906, as in 1903, a committee was established to review van der Heide's proposals.[31] Once again, there appears to be no surviving record of the committee's discussions, but it is clear that it advised postponement of the decision on the irrigation scheme for at least two years. In the meantime, the Canal Department was to continue with its limited programme of canal restoration and lock construction.

There is no record of the government undertaking a re-examination of van der Heide's major proposals in 1908. Instead, van der Heide himself came forward with two further schemes, one for the Pāsak area on the east bank, which was designed largely to improve the supply of irrigation water into the Rangsit canal system, and the other on the west bank.[32] Each project would cost 3 million baht, and would form an integral part of the originally proposed 'big scheme'. Yet even such relatively modest proposals found little favour within the administration. In a memorandum written in August 1908, the Financial Adviser, W. J. F. Williamson, argued that:

> To my mind it has not yet been satisfactorily shown that new irrigation works are required in Siam, except as feeders to already existing systems, owing to the want of sufficiently dense population, and I have consequently always been opposed to the Government committing itself to any of Mr. van der Heide's ambitious projects. The policy hitherto followed, of confining the energies of the Irrigation Department to the improvement of canals already in existence, has, therefore, commended itself to me, and I see no reason for any departure therefrom in the present circumstances of the country.[33]

Neither of the 1908 proposals was sanctioned.

The repeated refusal of the Siamese Government to accept any of van der Heide's substantial projects was undoubtedly a source of great personal sadness, as well as professional frustration, for the Director. Yet there is unlikely to have been much sympathy for him from within the government, for there is evidence that van der Heide commanded very little respect in the administration, even within his own department. By far the most detailed extant criticism of the Dutch engineer is contained in a report drawn up in 1910 by Mọm Anuruthathēwā, his successor as Director of the Canal Department.[34] Mọm Anuruthathēwā first condemned van der Heide as an incompetent administrator. It was usual practice for the Director to give instructions to his officials and to receive reports from them by word-of-mouth; consequently, there were few official records of the detailed work of the Department. Furthermore, there was no distinct demarcation of the specific duties of each official, such that it became difficult to ascertain the precise responsibilities of any particular individual. In addition, van der Heide had apparently tolerated a situation in which the Department's engineers, working independently, had sought to conceal the details of their work from each other—presumably in order to increase the opportunities for misappropriation and graft. In summary, there was no one with an accurate overall view of the full range of work in the Canal Department.

According to Mọm Anuruthathēwā, the financial administration of the Canal Department under van der Heide was similarly ill-disciplined. The Accounts Division of the Department had no function in the disbursement of funds; rather, senior officials had full authority over their own expenditures. The Department's cash resources were held not by its accountant but by van der Heide himself. There was no central division within the Department for the purchase of equipment and supplies, and consequently, there was considerable duplication of orders. Mǫm Anuruthathēwā was also highly critical of van der Heide's apparent inability to frame his irrigation proposals within the financial constraints of the government. He could think almost solely in terms of large-scale projects, demanding huge investments; he was not willing first to prepare modest proposals, involving a limited call on government funds, in order to demonstrate within a short period the anticipated benefits of irrigation.

Perhaps most seriously, Mǫm Anuruthathēwā accused van der Heide of professional incompetence. He alleged, for example, that on a number of occasions the Director had organized the construction of locks without making provision for the sluices through which excess water in the canals could be drained off. Consequently, during the heavy monsoon of 1909 the Phāsī Čharoen and Damnoen Saduak canals had seriously overflowed into adjoining rice, pepper, and garlic fields, causing substantial losses. More generally, Mǫm Anuruthathēwā argued that although van der Heide could be said to have shown considerable expertise in drawing up irrigation schemes, he had little authority as a practical construction engineer. This was a particularly serious disadvantage, in that van der Heide's long experience of irrigation work in Java frequently proved to be misleading in the very different geological and topographical conditions of Central Siam. Thus, Mǫm Anuruthathēwā argued, soil conditions in the alluvial delta of the Čhaophrayā were such that a considerable number of test bores had to be carried out before construction of a lock could begin; but van der Heide almost invariably neglected this preliminary survey, and simply employed the construction methods with which he was familiar from his work in Java. Consequently, many of the locks built under van der Heide's administration soon shifted their foundations and buckled. Finally, Mǫm Anuruthathēwā alleged that the inspections carried out during construction were usually perfunctory; that the majority of European engineers employed in the Canal Department had only recently graduated; that indolent on-site engineers would allow their labourers to idle away their working hours as they pleased; and that some engineers, permitted by van der Heide's lax financial administration to hire labour and purchase equipment and materials on their own initiative and without apparent restriction, had drawn substantial commissions from contractors and suppliers. The result of all these objectionable practices was that the quality of the construction work being carried out under van der Heide's administration was poor, whilst the expenditure was high. In view of these serious shortcomings—in administrative organization, financial discipline, and technical competence—it is not surprising that, again according to Mǫm Anuruthathēwā, the government's confidence in van der Heide had rapidly evaporated, leaving the administration quite unwilling to sanction any of his major proposals.

Mǫm Anuruthathēwā's testimony must be treated with caution. Van der Heide was initially appointed on a three-year contract,[35] and had that appointment been as disastrous as was suggested in Mǫm Anuruthathēwā's report, it is very unlikely that he would have retained his position in Siam for the length of time that he did.[36] In addition, it must be noted that Mǫm Anuruthathēwā was writing within just a few months of van der Heide's departure from Siam and as an official who, having served under van der Heide, now found himself appointed his successor; he may thus have felt it necessary to justify his recent elevation in part by denigrating the Dutchman. Moreover, the fact that Mǫm Anuruthathēwā referred in his report to attempts he made during the administration of van der Heide to reform some of the offending practices within the Canal Department, and that these attempts were strongly opposed by both European and Siamese officials, suggests that there may have been, as a result, considerable ill feeling between the two men.[37]

Nevertheless, there is evidence from other sources to support at least some of Mǫm Anuruthathēwā's allegations. Attached to his report was a memorandum by the Assistant Director-General for Constructional Works, a European, which laid out in detail the serious problems that had been encountered in the lock-construction programme during van der Heide's administration. Then in 1911, a series of investigations was carried out on a lock at Bāng Hia which had been constructed under van der Heide's authority but which had since buckled.[38] The investigations concluded that the foundations of the lock were insufficiently firm to take the pressures imposed on them; that insufficient ground tests had been carried out prior to construction; and that the design of the lock itself was defective in a number of important respects. Although it is possible that van der Heide was not directly to blame for these technical deficiencies, as Director of the Canal Department he was ultimately responsible for them.[39] Van der Heide's professional reputation may also have suffered in the eyes of the Siamese administration as a result of a lengthy dispute which he had with L. R. de la Mahotière, a Frenchman who had been engaged in the early 1900s to design a freshwater supply system for the capital. In developing his proposals, de la Mahotière strongly criticized van der Heide's irrigation scheme, partly on the grounds that it was too advanced for Siam's current requirements—agricultural Siam faced not a shortage of water but a shortage of labour to make productive the delta's extensive uncultivated tracts—and partly on the grounds that it would seriously disturb other important forms of economic activity in Central Siam.[40] In this last respect, he was particularly concerned at the possible impact of the irrigation scheme on the floating out of the teak logs from the forests of northern Siam, and on the work of the rice-mills and sawmills which lined the banks of the Čhaophrayā in its lower reaches. Van der Heide aggressively dismissed these criticisms,[41] and it was his rather disagreeable response, if not the actual criticisms themselves, which would have done little to enhance the Dutchman's reputation.

With regard to the financial administration of the Canal Department under van der Heide, the only additional documentary evidence concerns a long-running feud between the Director and the Minister of Agriculture that arose from a requirement that the Department draw its funds from its parent Ministry instead of directly from the Treasury as other departments in the administration were permitted to do.[42] This arrangement caused frequent cash shortages in the Canal Department, such that on occasions van der Heide felt compelled to draw on his personal account at the Hongkong and Shanghai Bank to keep the work of his engineers on schedule. By resorting to the use of private funds in this way, even for the most commendable reasons, it was undoubtedly more difficult for the Director to maintain strict financial accountability within his Department. With regard to the initial problem—the institution of an arrangement which led to an uncertain and inadequate flow of funds into the Canal Department—it is impossible to provide a clear explanation. It may have represented an attempt by the Ministry of Finance, and also possibly the Ministry of Agriculture, to frustrate van der Heide's major proposals; alternatively, it may have been a response by the Ministry of Finance to earlier extravagance on the part of the Canal Department.

Finally, there is evidence that van der Heide's personal manner could, on occasions, cause considerable offence to senior members of the Siamese administration. In negotiations with Čhaophrayā Thēwēt in 1903 over the terms of service for subordinate Europeans employed in the Canal Department,[43] and in his later dispute with the Minister over the interruptions in the flow of funds from the Treasury to his Department, the Director was disconcertingly forceful and impatient. His irritable dismissal of de la Mahotière's criticisms has already been noted. In addition, a later reference in the *Bangkok Times* that van der Heide 'alone [laid] down the law as to what Siam wants'[44] would imply a distinctly imperious character. Moreover, van der Heide showed little hesitation in pressing on the Siamese administration his views on wider political issues, for as he explained in a letter to Čhaophrayā Thēwēt in May 1907, he regarded himself a political economist as well as an irrigation engineer.[45] Attached to that letter was a lengthy memorandum, 'Views on Practical Politics for Siam', in which van der Heide condemned, mainly on political grounds, the government's railway construction programme. He pointed out that the completion of the southern railway linking Bangkok and the Malay States, and the completion of the French line from Hanoi to the Mekong, would make it possible to travel from the Bay of Bengal to Hanoi by rail, with the exception of the short distance from Kōrāt on the northeast plateau of Siam to the Mekong border. The French, suggested van der Heide, would soon put considerable pressure on the Siamese Government to construct that short line—in time, France would almost certainly move against Siam herself, for the commercial community and 'still more the military authorities of the French colony will object to having the nearest line of communication with France left in the hands of a weak Asiatic state'. The Siamese ministers are unlikely to have appreciated this rather alarmist analysis from a man engaged as an authority on irrigation. Finally, reference must be made to a highly critical assessment of van der Heide by Chulalongkorn himself, made on the eve of the former's departure from the Siamese service.[46] The King scathingly commented that under van der Heide, the Canal Department had failed to build a firmly structured programme of work, but had continuously let schemes fall abandoned, half pursued. Van der Heide himself was simply ill disciplined—he would accomplish nothing of substance in Siam. 'It would be good to let him go.'

In early 1909 the Siamese administration, under pressure to restrain its fast increasing expenditure, finally decided to abandon all proposals for large-scale irrigation works.[47] This decision was reached at a meeting in March attended by Prince Čhanthaburī (the Minister of Finance), Čhaophrayā Thēwēt, W. J. F. Williamson, and van der Heide. All survey work for the 'big scheme' was to cease, and over the following decade government irrigation work was to involve simply the completion of the few minor projects already underway, and the maintenance of existing facilities. Van der Heide was asked to prepare a schedule of work for the Canal Department within this brief, and to make precise recommendations for a very major reduction in the Department's staff.[48] His first recommendation was for his own dismissal.

Van der Heide reacted bitterly to these events, blaming Prince Čhanthaburī for what would be, in effect, the dismantling of the Canal Department.[49] Tempers flared towards the end of April when the Ministry of Finance refused to disburse further funds to the Department, on the grounds that receipts for earlier withdrawals were still outstanding. On this occasion, van der Heide was apparently not prepared to resort to his personal account at the Hongkong and Shanghai Bank to maintain the work of the Department, for he immediately ordered all work to cease.[50] Two senior officials in the Ministry of Agriculture succeeded in restoring the Canal Department's funds and, apparently, in soothing van der Heide's injured feelings, but it was now clear that the Director was eager to get away from Siam at the earliest possible opportunity. He sought leave of absence with effect from 10 May but his Ministry, arguing that the restructuring of the Canal Department's work must first be completed, refused the request.[51] Van der Heide eventually left Bangkok on 13 June 1909, seven years to the day from his arrival in the kingdom, to resume his career in the irrigation service of the Netherlands East Indies.[52]

III

For the five years following the departure of van der Heide, the reduction in the Canal Department's expenditure budget, although substantial, was not as severe as had been proposed in 1909.[53] But, in fact, more significant than the magnitude of the Department's budget in this period was a change in its orientation away from irrigation work and towards the improvement of inland water communications. This change was made explicit in April 1912 when the Canal Department was removed from the Ministry of

Agriculture, placed in the Ministry of Communications, and renamed the Department of Ways.[54] In this reorganization the Ministry of Agriculture was to retain responsibility solely for planning those canal projects that would primarily assist cultivation. There was a clear implication that such irrigation work would be on a very limited scale.

Yet, less than a year after the transfer of the Department to the Ministry of Communications, the administration was again giving serious consideration to the implementation of a large-scale irrigation programme. It is usually argued that this revival of interest was prompted by three years of abnormal rainfall from 1909/10—one year of high flood followed by two years of drought.[55] Undoubtedly this was an important contributory factor, but there were other significant considerations. The case for a major irrigation initiative appears to have been revived by a new Minister of Agriculture, Prince Rātburīdirēkrit, who was appointed, ironically, on the very day the Canal Department was removed from his Ministry.[56] In a letter to the King in January 1913, Prince Rātburī argued that the construction of major irrigation works was essential if the kingdom's vitally important rice trade was to maintain its competitive position against those of Lower Burma and Cochin-China.[57] He suggested that the British administration in Burma had already undertaken a major programme of irrigation works and that further projects were being planned, whilst in Indo-China, it was even being proposed to abandon construction of several railway lines and to redeploy the capital in irrigation projects. As a result of these investments in irrigation, Prince Rātburī argued, the cost of rice cultivation in the two European colonies was being held stable, whilst in contrast in Siam it was continuing to rise. 'Before long', he concluded,

Siamese rice will be unable to compete against rice from Burma and Cochin-China, and the kingdom will become increasingly impoverished. If Siam is to continue to export rice, it is essential that the population is able to cultivate the crop in conditions as favourable as those enjoyed by cultivators in other countries.[58]

Prince Rātburī then surveyed the history of canal excavation in the kingdom from the time of the Siam Land, Canals, and Irrigation Company in the 1890s in an attempt to draw lessons that would guide the administration as it formed its new proposals. He argued, for example, that the company, and then in the 1900s, the

Canal Department, had been concerned primarily with opening up new land for agricultural production, and had done relatively little to facilitate the actual cultivation of rice. In other words, both had been engaged essentially in canal cutting, not true irrigation work. Yet there was a clear need in Siam, the Minister continued, for a distinct irrigation programme in the full sense.

Not unexpectedly, Prince Rātburī was critical of many aspects of van der Heide's work in Siam. It is interesting to note, however, that he was also critical of his government's handling of the initial appointment of an irrigation engineer, as well as its subsequent treatment of van der Heide himself. Specifically he argued that in selecting van der Heide, the government had appointed an engineer who had no experience of, or expertise in, the type of soil conditions and topographical structures to be found in Central Siam. Moreover, on his arrival in the kingdom, the government had failed to specify its irrigation requirements and intentions, but had left van der Heide to find his own way without guidance or direction. It was therefore not surprising that the government had failed to secure an adequate return from its substantial investment in the Dutch engineer's appointment, although, Prince Rātburī concluded, much of van der Heide's work would still have value for the administration as it embarked on a new irrigation initiative.

The Minister of Agriculture further pointed out that whereas during the 1900s the work of the Canal Department had been financed solely from the government's tax revenues, the irrigation programme which he would propose could be financed only by raising a foreign loan. This procedure could bring very considerable risks, for if the capital so borrowed were misapplied and thus secured no benefit for cultivation in Siam, the international financial standing of the kingdom would be seriously undermined—indeed, Siam's political independence could be threatened.

Prince Rātburī was adamant that the government should not simply drift into an irrigation scheme as had allegedly occurred with the appointment of van der Heide. Rather, the decision on an irrigation initiative must now be irrevocable, and the commitment to securing the required capital total. Without such a firm undertaking from the government at the outset, it would be impossible for the administration to attract a first-rate irrigation engineer to Siam. At this point, the Minister of Agriculture was not able to provide a detailed outline of the proposed scheme, beyond that it would necessitate a weir across the Chaophrayā River, and that it would require an investment of between £2 million and £3 million.

In short, the case for a large-scale irrigation scheme as presented by Prince Rātburī was that it was essential for the maintenance of Siam's commercial prosperity, and thus for the protection of the economic welfare of her people. His proposals were briefly discussed and then approved at a meeting of the Council of Ministers on 20 January 1913, and one month later the Minister received permission from the King to proceed.[59] Almost immediately an invitation was sent to Thomas Ward, a superintending engineer in the irrigation service in the Punjab, to prepare an irrigation scheme for the Čhaophrayā plain that would cost no more than £1.75 million.[60] He arrived in Bangkok in September 1913. In October the following year, the canal divisions recently transferred to the Ministry of Communications were returned to the Ministry of Agriculture and reconstituted as the Irrigation Department.[61] The Department's Director was R. C. R. Wilson, also from the Punjab service and one of Ward's assistants.

Thomas Ward's report was completed in February 1915.[62] In an attached minute, Prince Rātburī again stated that the government's aim in undertaking a major irrigation programme would be 'to enable the farmers of Siam to maintain against the increasing competition of neighbouring rice-growing states, fostered by energetic governments, the position hitherto held by Siam in the rice markets of the world'.[63] Ward proposed the implementation of five projects, the most important being the Suphan scheme to the west of the Čhaophrayā River, to which he gave priority, and the Pāsak South scheme on the east bank. The total cost of all five projects was 22.75 million baht which, at the current rate of exchange, was exactly equivalent to the £1.75 million stipulated earlier by the government as the sum it was prepared to invest in irrigation. It was estimated that it would take 6–8 years to complete the construction work. With regard to the proposal that a weir be constructed across the Čhaophrayā, Ward argued that

> ... such a barrage would increase the capital outlay on irrigation by some 4½ m. baht ... and it would affect a larger tract of land than the population of the country could tackle effectively at present; a tract which would, moreover, demand immediate development to provide adequate interest on the capital invested.... Therefore construction of such a work under present conditions would be a mistake. But the time may well come when, with a strong demand from a growing population for large new tracts of irrigated land, ... the Government will find a main river barrage

not only desirable but necessary. By such time the Irrigation Department ... will have had time to construct a complete system of irrigation and drainage channels to absorb at once the supplies from the barrage and to turn the same to immediate profit ... and the population will have had time to grow accustomed to far-reaching changes in the conditions of agriculture and of life generally, and will be ready and anxious to adapt itself to the still greater changes which a main river barrage must bring about.[64]

In brief, Thomas Ward's proposals were relatively modest, seeking simply to meet the perceived current irrigation requirements of the kingdom. In this respect they contrasted sharply with van der Heide's strikingly more ambitious proposals of 1903.

But there was an ironic similarity in the circumstances in which the 1903 and 1915 reports were completed and then presented to the government. In the earlier case, the major acceleration of the government's railway construction programme in the aftermath of the Shan uprising—a development which occurred whilst van der Heide was writing his report—severely disturbed the administration's earlier commitment to a major irrigation initiative. In the case of the Ward report, a comparable disruption of the government's intentions arose from the outbreak of war in Europe in August 1914.

In February 1915, the same month as Ward was completing his report, the Minister of Finance, Prince Čhanthaburī, brought to the attention of the King a serious problem that had arisen in the construction of the government's expenditure estimates for 1915/16.[65] He reminded Vajiravudh that it had been the government's intention to raise a loan of £2.75 million in Europe during that year, £1 million to be used to complete the northern railway to Chiangmai, the remainder being for the irrigation scheme. Now, with the outbreak of war in Europe, there could be no recourse to the European money markets for several years. Furthermore, the proceeds of earlier loans were nearing exhaustion, and the Treasury reserve was relatively low. In these difficult circumstances, Prince Čhanthaburī stated, a decision would now have to be made between the irrigation and railway programmes.

In fact, the Minister of Finance argued forcefully that preference should be given to the irrigation proposals. The construction of the final stretches of the northern railway to Chiangmai, through hilly terrain, would present considerable engineering problems. In addition, because of the war in Europe, it was becoming difficult to hire sufficient railway engineers, and impossible to import railway materials from Germany as had long been the practice. Finally, the completed line from Lampāng to Chiangmai would run through economically less important territory, and was therefore unlikely to be notably remunerative. In contrast, argued Prince Čhanthaburī, the return from a major irrigation investment—in terms of an expansion in rice production—was certain to be considerable. Indeed, the Minister urged that construction work should begin almost immediately. The government had already invested a considerable sum in the preparatory work under Ward, whilst to postpone further might well involve the administration in additional initial expenditure at a later date. Perhaps most importantly, with every day that passed, the competition from neighbouring states in the international rice market grew more intense.

Prince Čhanthaburī's proposal that construction of the northern railway be halted did not imply that funds would thereby be released for Ward's irrigation projects for, as noted earlier, both the railway and irrigation programmes were originally to have depended on a foreign loan. To meet the construction costs of irrigation works, the Minister of Finance now suggested that the government should raise a loan within Siam herself. Although Prince Čhanthaburī's proposals were approved in full by Vajiravudh,[66] the Minister of Communications (and former Minister of Agriculture), Čhaophrayā Wongsānupraphat, reacted angrily to the argument that construction of the northern railway be halted, even if only temporarily, and a series of reports appeared in the Bangkok press in early March 1915 that he was pressing for completion of the line to Chiangmai without delay.[67] Prince Čhanthaburī was furious that this controversy should be aired in public, but it would appear that the Minister of Communications was successful in maintaining progress on the construction of the northern line.[68] Necessarily this made more difficult official acceptance of Ward's irrigation proposals.

Thomas Ward's report was formally presented to the King by Prince Rātburī at the end of April 1915.[69] The Minister of Agriculture was in sombre mood. He pointed out that the war was not only making it impossible for Siam to raise a foreign loan, but also causing an acute shortage of European engineers in the East as large numbers of men returned home to enlist. Indeed it was now impossible for Siam to find competent irrigation engineers, despite the fact that the government was offering salaries three times the level being paid in British India. Furthermore, Prince Rātburī was clearly displeased with the attitude of those few European engineers who had remained in the Siamese service, for predictably they had taken advantage of the current crisis to demand more favourable terms. To make matters worse, the Director of the Irrigation Department himself, R. C. R. Wilson, appeared to the Minister to be, in some way, unsuited to his position, and he therefore could not now be expected to stay in the kingdom for long. But then, finding a suitable replacement for this crucial post would, in the circumstances, be extremely difficult. The only source of comfort for Prince Rātburī in this respect was the recent emergence from within the Irrigation Department of one particularly able Siamese official: but even he would require more time, building expertise and experience, before he could take over the senior positions.

In April 1915, the Minister of Agriculture offered detailed comment on only one aspect of Ward's specific proposals. This concerned his calculation that, partly through the imposition of an irrigation tax, the government's tax revenue from agricultural production and trade following the completion of the irrigation works would be almost three times its current level. Prince Rātburī seriously doubted whether such an increase was in any way realistic. More particularly, he questioned whether Ward had the specific skills to undertake such specialist calculations, and he suggested that the administration's revenue officials be asked to re-examine the figures.

However, the critical issue was, of course, to secure the resources to finance the new proposals. In writing to Vajiravudh in April 1915, Prince Rātburī made no reference to the raising of a loan within Siam as had been suggested two months earlier by Prince Čhanthaburī; and in fact the Minister of Finance himself was now proposing an allocation for irrigation on a year-to-year basis. As it was quite clear that that allocation would fall far short of the sum required to carry out the 1915 irrigation report in full (not least because the Ministry of Finance was now also required to finance continued construction of the northern railway), a decision would now have to be made between the specific projects drawn up by Ward. Prince Rātburī's own view was that the Pāsak scheme appeared to offer the greatest possibilities.

Prince Rātburī's report and proposals appear to have stimulated relatively little interest on the part of Vajiravudh, for he merely referred the matter to Prince Čhanthaburī, 'in case he has any ideas that could ease the situation'.[70] Indeed it would appear that the fate of Ward's report was decided essentially in discussions between Prince Čhanthaburī and Prince Rātburī over the middle months of 1915.[71] The two Ministers first confirmed that the simultaneous implementation of all Ward's projects, as Ward had proposed, would not be possible—or indeed prudent. In this last respect, Prince Rātburī was particularly concerned that the administration still was unable to ascertain with any accuracy the increased revenues which it might anticipate from the schemes under review (he remained sceptical of Ward's optimistic calculations),[72] and was thus concerned that the government, once committed to the full implementation of the Ward report, should then discover that it could not recover its investment costs. Prince Rātburī and Prince Čhanthaburī therefore proposed that the Irrigation Department's priority lay in completing and repairing existing works, particularly those in the Rangsit area, responsibility for the area having reverted to the government in 1914 on the expiration of the company's concession. With regard to the individual projects set out by Ward, the Ministers of Agriculture and Finance confirmed Prince Rātburī's earlier suggestion that the government should first proceed with just the Pāsak scheme. In the absence of a foreign loan, work on this scheme could be financed from the Treasury reserves.

As was noted earlier, Ward in fact had argued that priority be given to the Suphan project, but there is no indication in the present documents as to the reasons why that recommendation was overturned. It could not have been on the grounds of cost, for the Pāsak scheme was anticipated to require substantially more funds than the Suphan scheme. However, in a published report of the Royal Irrigation Department in 1927, it was suggested that the decision to proceed first with the Pāsak project had been taken 'probably because it was considered inadvisable to disturb existing arrangements of landlord and tenant in the Rangsit area and elsewhere, which the opening up of big areas of land in Suphan, free to all, must have done'.[73] A related interpretation was offered by the Financial Adviser, Williamson, who argued that a major influence in the decision was the fact that the Pāsak project, to the east of the Čhaophrayā, would improve irrigation conditions in an already populated area, in contrast to the Suphan project which would involve opening up a new, relatively unpopulated, area on the west bank.[74]

Work on the Pāsak project began in November 1916,[75] and

within two years the annual budget allocation for irrigation, drawn from the Treasury reserves, was running at over 2.5 million baht.[76] Even so, progress was relatively slow, for the war in Europe made it extremely difficult to obtain many essential materials from abroad, or ensured that they could be obtained only at vastly inflated prices.[77] The Pāsak project was eventually completed in 1922, at a cost of almost 15.5 million baht, compared to the 11.5 million baht originally estimated by Ward.[78] Work then began on the Suphan project, but once again construction proceeded slowly, budget stringency in the later 1920s and then the collapse of rice prices during the depression years both acting to retard the work. In these ways, the construction of irrigation facilities in Siam made only limited progress in the period down to the 1930s.

IV

The policy of the Siamese administration of the late nineteenth and early twentieth centuries towards the construction of large-scale irrigation facilities in the kingdom and, in particular, its responses to the proposals presented by van der Heide in 1903 and by Ward in 1915, have already been considered at length in a number of English-language studies. It is proposed to review briefly the main conclusions of some of those earlier studies, before turning to reconsider the evidence which has been presented in the preceding sections of this chapter.

In his *Economic Change in Thailand Since 1850*, first published in 1955,[79] James C. Ingram offered four principal reasons for the failure of the administration to undertake major irrigation works in the early twentieth century.[80] First he argued that, as the government felt that it was not practical to charge irrigation fees to the rice cultivators of the Central Plain (for the farmers there were not accustomed to paying for the water which flowed into their fields from the region's distribution canals), major irrigation works could not easily be made to pay for themselves directly. In turn, this made irrigation 'less attractive to the government than projects such as railways, electricity, and water works, which did promise a direct return'. Second, the benefits of irrigation in Siam would not have been as overtly spectacular as they were, for example, in Egypt where desert was transformed into fertile land. The benefit of irrigation in Siam was simply that it would lessen the risk of crop failure arising from seasonal and yearly variation in rainfall, and this implied that in years of good rains, irrigation would secure little additional benefit for the rice cultivator. Third, Ingram noted that contemporary opponents of a major irrigation initiative had argued that the kingdom's population was too low to ensure cultivation of all land newly opened by large-scale irrigation projects. And finally, he drew attention to the fact that the resources of tax revenue and foreign capital available to the Siamese Government during this period were limited. This implied that certain proposed public works projects had to be forgone, and 'because there were pressing political as well as economic arguments for railway construction, irrigation was [thus] consistently postponed'. Ingram continued:

> It was in this context that the influence of the Financial Adviser was important; ... in view of the facts that revenue was limited partly by foreign treaties, that British Financial Advisers largely decided whether additional foreign loans were wise or not, and that Williamson obviously did not approve of irrigation in Thailand, a large measure of responsibility for the delay must be borne by him and by Britain, the dominant power in Thailand both politically and economically.[81]

Among more recent scholars, David Feeny has also drawn attention to the strong opposition of the British Financial Advisers to proposals for large-scale irrigation projects.[82] Indeed, in the context of Williamson's opposition to irrigation, and commensurate support for a substantial investment in railway construction, Feeny argued that important British interests were at stake. With regard to the latter, a significant proportion of the rails and rolling stock, as well as consultancy services were purchased from Britain; whilst increased rice productivity in Thailand following the completion of a major irrigation programme 'might have provided more effective competition [for] Burma, a British colony and major rice exporter'.

However, the main body of Feeny's explanation for the failure of the Siamese Government to undertake major irrigation projects in this period was contained in three propositions which focused in large part on the personal political and economic interests of the Siamese administrative élite itself.[83] First, on the assumption that the principal competitor of irrigation for the government's scarce resources was indeed railway construction, he argued that the latter took precedence because it provided political security and public administration benefits to the élite and the kingdom, whilst irrigation would have provided primarily economic benefits. Second,

Feeny argued that there was substantial opposition to major irrigation projects in the Central Plain from the landowners at Rangsit, who feared that the creation of new, effectively irrigated districts would draw tenants away from their land, and so lead to considerable falls in rents and land prices in the concession. It was for this reason, he suggested, that in 1915 the decision was made to proceed with the Pāsak scheme, rather than the Suphan project, for the former would have improved irrigation conditions in Rangsit and so have led to an appreciation in land values there, whilst the latter may well have provoked a damaging exodus of cultivators from that district. The corner-stone of this argument was that among the principal landowners in Rangsit were many powerful members of the Bangkok administration. This leads to Feeny's final proposition, that 'the government and its officials had only limited means by which they could appropriate some of the gains from public investments in irrigation'. The argument here was twofold. It was suggested that those members of the Bangkok administrative élite who had invested heavily in land at Rangsit from the late 1880s, were probably not in a financial position to extend their landholdings substantially into any new area opened by a major public irrigation initiative. Indeed, as noted above, such an initiative may well have actually damaged the value of their existing investments in the Rangsit district. Second, because Siam did not have fiscal autonomy until 1926, the administration could not raise land and foreign trade taxes effectively to capture a major proportion of the increased productivity in rice cultivation which would arise from investment in irrigation. Lack of fiscal autonomy also limited the resources available to the administration for investment and the repayment of foreign loans. Feeny summarized his arguments as follows:

Railways provided more in national security, political development, and public administration benefits than irrigation, and because the capital budget of the government was particularly limited, little was left over after military and public administration projects were pursued. The interests of the landlords at Rangsit and the Bangkok élite favoured railway over irrigation investments, despite the fact that irrigation investments were more beneficial than railways to the society as a whole. The divergence between the rate of return to the society and the rate of return to the policy makers, as well as the divergence between the economic rate of return and the over-riding goal of maintaining national independence explain a great deal about the failure of Thailand to develop through rice exporting.[84]

From the analysis of Leslie E. Small, two arguments are particularly worthy of emphasis.[85] Small argued that in the early twentieth century, the Siamese administration primarily required of irrigation projects that they 'stabilize' rice production in areas already under relatively intensive cultivation. There was little government interest in proposals, such as those of van der Heide, that envisaged the rapid opening of large, new tracts for the cultivation of rice, for it was feared that this process would produce 'undesirable political, social, and economic consequences'. Second, he argued that the failure of the Siamese administration of this period to mobilize directly either the labour or the financial resources of the rice cultivators themselves to meet part of the construction and maintenance costs of irrigation works, threw the full burden of that investment on to the state. Inevitably, this acted to discourage a major irrigation initiative.

Finally in this brief survey of earlier studies, reference should be made to David Johnston's observation that one aspect of the rejection of van der Heide's recommendations 'no doubt relates to the inner political workings of the government at the time'.[86] Specifically, he referred to the failure of Čhaophrayā Thēwēt to argue persuasively for irrigation in 1903, and to van der Heide's apparently volatile relations both with crucial figures in the administration and with his own Department. However, Johnston also suggested that the major obstacle to the progress of irrigation in this period appears to have been the demands of railway construction.

Not all the arguments summarized above would find supporting evidence in the earlier sections of this chapter. For example, although it is clear that the Financial Advisers, notably W. J. F. Williamson, did indeed strongly oppose van der Heide's proposals, there is no evidence that that opposition influenced the decisions to defer and then reject the Dutchman's schemes—if by 'influence' is meant changing the established opinion of the administration, or asserting views which bring it to a decision. The evidence is clear that the Siamese ministers formulated their own firmly-held opinion of van der Heide's proposals, and took the actual decisions on irrigation within their own counsel. Williamson merely agreed with their assessment, and applauded their actions.[87] Moreover, Feeny's reference to Britain's substantial interest in supplying railway materials to the kingdom, and to a possibly detrimental impact on Burma's rice export trade were Siam to proceed with a large-scale irrigation programme, should not, as it stands, be taken as evidence of Williamson's motivation in preferring railway to irrigation construction. In any event, the Siamese administration was invariably quick to spot and curb the national self-interests of its European employees.

A number of reservations must also be noted with respect to the argument that the Siamese administration was either unwilling or unable to consider the imposition of charges on the cultivator for the supply of irrigation water and for the provision of newly-opened land, measures which would have secured for the government a substantial direct return from a major investment in irrigation construction. In fact, such charges were proposed by both van der Heide and Ward,[88] and by successive Ministers of Agriculture. In November 1899, Chaophrayā Thēwēt envisaged the sale of land on the banks of canals newly-cut by the government;[89] van der Heide's 1903 report contained similar provisions for the sale of land, and for the introduction of a water tax of one baht per *rai*; and in 1915 Ward proposed, with the apparent support of Prince Rātburī, both the imposition of a water tax, and the introduction of a charge on cultivators where the Irrigation Department undertook the construction of local irrigation ditches.[90]

It is difficult to provide an adequate explanation for the firm reluctance of the Siamese administration as a whole to sanction such charges, and thereby make irrigation pay for itself directly. Ingram has suggested that in fact the sale of land and water in this manner was contrary to traditional practice in Siam;[91] but against this may be set the advocacy of such measures by the government's own Agriculture Ministers as noted above, as well as the fact that in the later 1870s the administration had, in effect, sold newly-opened land to cultivators, when it had required land applicants to bear part of the costs of the canal excavations in their district.[92] A related suggestion by Ingram, that van der Heide 'may have misjudged the willingness of farmers in the Central Plain to pay money for water and land',[93] must also be treated with some caution, for no evidence is offered for an attitude which, in circumstances where irrigation would indeed have secured substantial advantages for the rice cultivator, would run counter to the well-established economic responsiveness and discrimination of the Siamese peasant farmer. Finally, according to Small, W. J. F. Williamson opposed Ward's proposed water tax on the grounds that it would constitute an additional financial burden on the landowners and cultivators which they would find difficult to bear.⁹⁴ However, as the only landowners and cultivators to face this charge would be those whose income was, presumably, being raised through the introduction of irrigation facilities, it is difficult to see the force of the Financial Adviser's argument.

In fact, the Siamese administration of this period may well have anticipated that the principal return on an investment in irrigation would be secured not through the imposition of direct charges but indirectly—through the increased yields from the taxation of agricultural land, from customs duties, and from the major tax-farms (for opium, liquor, and gambling), which could be expected to follow an expansion in rice production and the concomitant rise in the general prosperity of the kingdom. Certainly, Thomas Ward anticipated that completion of major irrigation works would secure a very substantial increase in the government's revenue specifically from the taxation of agricultural land, and although Prince Rātburī may have been sceptical of Ward's actual calculations, both he and Prince Čhanthaburī appear to have accepted the principle which lay behind them.⁹⁵ However, it is Feeny's contention that the ability of the Siamese administration to raise a significantly larger revenue from agricultural land, and thus capture a major part of the gains from an irrigation investment, was severely constrained by the fact that the rates of land tax in the kingdom were frozen at a low level by the treaties concluded between Siam and the major powers from the middle of the nineteenth century.⁹⁶ In fact, the administration may have had greater flexibility in this respect than Feeny implies. First, under an agreement signed with Britain in September 1900, the schedule of land taxes attached to the Supplementary Agreement drawn up with Sir Harry Parkes in 1856 was abrogated, and in return Siam provided a new undertaking that the rates of land tax in the kingdom would not exceed those levied on similar land in Lower Burma.⁹⁷ In 1905, the administration used the provisions of this agreement to reform the kingdom's land tax system, restructuring and redefining the categories into which agricultural land would be assessed for the purposes of taxation, and increasing the tax rates themselves. The reform immediately doubled the government's revenue under this head, and although a major part of this particular increase was used simply to replace revenue lost by the closure of provincial gambling dens over the years 1905–7, the primary fact remains that from 1900 the Siamese Government had secured a considerably greater measure of autonomy in its taxation of agricultural land. A second moderating consideration in this respect was that the construction of irrigation works in a particular district would permit the government's revenue officials to place the land in that locality into a higher tax category, for irrigation should clearly have raised its productivity. In this way, the total assessment of a newly-irrigated district could be increased substantially, even while the structure of land tax rates remained unchanged.

Yet, the critical issue concerns not the inability or disinclination of the Siamese administration to make irrigation pay for itself, but whether that consideration was an important influence on the government's actual decisions to accept or reject the major irrigation proposals placed before it. What little evidence there is on this point relates primarily to the period of the Ward proposals. When Prince Rātburī first outlined his irrigation initiative to the King in January 1913, it was evident that the Minister was prepared to proceed with the foreign loan which he considered essential to finance a major programme, only because he anticipated that the government's financial return from irrigation would meet the interest and repayment charges.[98] His later unease with respect to the accuracy of Ward's estimates of land tax revenues reflected the importance of that assumption. It should also be noted that Ward himself opposed the construction of a weir across the Čhaophrayā principally because 'it would affect a larger tract of land than the population of the country could tackle effectively at present; a tract which would, moreover, demand immediate development *to provide adequate interest on the capital invested*'.[99] It might finally be added that when Prince Čhanthaburī was discussing the financing of Ward's proposals in early 1915, he referred to a stipulation laid down by the administration when it had first sought a European loan in the 1900s that capital so raised be employed only for projects which would secure a substantial financial return.[100] The Minister of Finance clearly accepted that the irrigation proposals then before the government met that important requirement.

Nevertheless, there is also evidence to suggest that *in practice* a concern with the anticipated return from irrigation construction exerted little influence as the Siamese administration determined its irrigation policy. The fact that the administration was firmly unwilling to impose the direct charges on the cultivator which would have provided it with an easily assessed and reliable revenue, and the absence of a compelling explanation for the government's reluctance in this matter, suggests that the need to secure a substantial return on irrigation investments may not, after all, have been of major importance. Second, the stipulation referred to by Prince Čhanthaburī that only major revenue-creating projects be financed by means of foreign loans, in practice appears not to have been rigidly followed. A major part of a £3 million loan raised in Europe in 1907 was used to build up Siam's sterling reserves, whilst loans raised in 1922 and 1924 for a total of £5 million were intended to restore the depleted reserves of the Treasury.[101] Finally, evidence presented earlier in this chapter suggests that the primary influence which determined the Siamese administration's response to Ward's proposals was not concern that the government should secure a firm return from its irrigation investment (despite Prince Rātburī's sceptical references to Ward's calculations), but rather the serious difficulties involved in securing sufficient resources to implement the report in the first place.

Of the other arguments presented by Ingram, Feeny, Small, and Johnston, the majority are either supported by the evidence presented earlier, or at least not seriously contradicted by it. However, simply to list the influences which shaped the Siamese administration's irrigation policy in the early twentieth century does not constitute a full explanation of its perceptions and decisions. That requires, at the very least, an evaluation of the relative importance of those various influences, and whether their relative importance changed over time. That task is a difficult one, essentially because the administration's irrigation debate is poorly documented; most seriously there appears to be no surviving record of the crucial deliberations of the ministerial committees established in 1903 and 1906. But it is a task that will occupy the remainder of this chapter.

V

On the basis of the evidence presented earlier I would argue that by far the most important reason for the rejection of van der Heide's main proposals of 1903 was their cost, in relation both to the volume of resources at the government's command and to the demands of the other major expenditure programmes before the administration.[102] The Dutch irrigation engineer proposed an investment of 47 million baht extended over twelve years, with a maximum expenditure in any one year of 5.5 million baht.[103] These expenditure proposals may be compared with the figures for the government's total revenue in 1903/4 of just under 43.5 million baht, and for the revenue from the administration's single most lucrative source, the opium farms, of a little short of 7 million baht.[104] Clearly, therefore, in relation to the revenue resources of the Siamese state, van der Heide was proposing a very substantial commitment to irrigation.

However, as earlier writers have indicated, the critical consideration here was less the size of the proposed investment in irrigation in relation to government revenue, but rather the status of irrigation in the administration's order of expenditure priorities.[105] More specifically, it has been argued that in a period when the very independence of Siam was seen to be threatened by the imperial European powers, the Siamese administration understandably gave clear preference to those measures which would serve to defend the kingdom's political sovereignty. In practice this involved, most notably, the construction of a railway network radiating from Bangkok to facilitate the effective administration of the whole kingdom from the capital, a strengthening of the armed forces, and the creation and maintenance of currency reserves sufficient to secure the external value of the baht. However, earlier writers have failed to convey the stark urgency with which the Siamese administration of the early twentieth century was forced to enter into these commitments. Thus the construction of railways and the strengthening and reform of the armed forces were seen as *absolutely essential* to the firm government of the kingdom from the capital and to the preservation of the kingdom's security,[106] whilst Siam's reputation in the powerful trading and financial centres of the world was seen to rest in large measure on the maintenance of currency stability and the pursuit of financial orthodoxy. In short, these were commitments which could not, in any circumstances, be avoided. In contrast, although the implementation of a large-scale irrigation programme in the Central Plain would have secured a major expansion of the rice economy, without it Siam was still to develop into one of the world's principal producers and exporters of rice. Irrigation may have been very desirable, but it was not absolutely critical to the maintenance of the established order in early twentieth-century Siam. There is a further point. If the Siamese administration's investments in railway construction, the military, and exchange reserves were to be effective, then they had to be on a substantial scale. As the Shan attack on Phrāe in mid-1902 demonstrated, limited construction of the northern railway only as far as Lopburī did little to secure Bangkok's firm control of the provinces in the north.[107] Similarly, the government's exchange reserves had to be of a minimum (and substantial) volume if they were to discharge their function effectively, that is to maintain the external value of the baht at par in all normal trading conditions.[108]

However, this analysis does not explain the rejection of van der Heide's later, less ambitious schemes, particularly his final proposals of 1908 which were estimated to cost in full only 6 million baht.[109] Neither does it explain why a scheme for the Pāsak area drawn up by van der Heide in 1908 and costing only 3 million baht was rejected, whilst in 1915 a further scheme for the same area, but this time estimated to cost 11.5 million baht, was sanctioned. These specific questions can be approached most effectively through consideration of a more general proposition. To what extent did the Siamese administrative élite of this period appreciate the benefits which an appropriately designed irrigation scheme would secure for the kingdom? And consequently, how firm was their enthusiasm for irrigation? From the evidence presented earlier it is clear that the administration's perceptions and ardour were not constant, but that there were substantial shifts in official appreciation and enthusiasm over this period. At the turn of the century, as the government began its search for an irrigation engineer, the King and Čhaophrayā Thēwēt each made clear his enthusiasm for a major initiative, although it would also be true to say that neither appeared able to state, even in the most general terms, the measures he envisaged or to outline with precision the objectives which a substantial irrigation programme was to secure. However, by the time of the submission of van der Heide's report in early 1903, that enthusiasm had evaporated as the government's financial position had deteriorated. Most dramatically, by August 1903, the Minister of Agriculture himself could be found sharply reducing the already much-reduced expenditure estimates prepared by van der Heide. This antagonism towards irrigation was to continue until the end of the decade.

It is in this context that van der Heide's occasionally abrasive relationship with senior members of the Siamese Government, his alleged maladministration of the Canal Department, and the criticisms of his professional competence, are significant. In 1903, the government was clearly unnerved by the ambitious scale and heavy financial demands of van der Heide's initial proposals. Consequently, in the years which followed, it became necessary for the case for a major irrigation initiative to be made (perhaps for the first time), by sober argument and by practical demonstration through the implementation of a limited number of modest, but highly effective, self-contained schemes. Because of his character, and in particular his near-obsession with 'the big scheme', van der Heide was not the person to make that case. Indeed it is possible that by the end of the decade, van der Heide's standing within the administration had deteriorated to such an extent that an objective assessment of the arguments for a major irrigation initiative by the government's ministers waited upon the Dutchman's departure from the kingdom.

In one important respect the Siamese ministers themselves were responsible for the disintegration of their relationship with van der Heide. At the time of the latter's initial appointment, in mid-1902, the administration had no distinct view of the kingdom's irrigation requirements and, more importantly, had made no decision as to the volume of funds it would be prepared to set aside for an irrigation programme. Consequently, as he came to write his major report in the final months of 1902, van der Heide may have had reason to believe that the administration had, in effect, given him full discretionary authority. Given these circumstances, his later frustration, indeed bafflement, in the face of the administration's protracted rejection of his main proposals is more understandable, for he could argue that he had simply given his Siamese employers what he understood they had requested—'technical reports about the *best* Irrigation systems for the Lower Provinces of Siam and estimates of their probable costs'.[110] In this context, it is surely significant that when the government engaged Thomas Ward in 1913 it firmly stipulated that he was to prepare an irrigation scheme that would cost no more than £1.75 million.

The reasons for the revival of interest in a major irrigation initiative in the early 1910s were considered at length earlier. Here, it need only be emphasized that in this later period the Siamese administration clearly had a more sophisticated appreciation of the need to construct major irrigation works in the kingdom and of the precise benefits they would secure. This is most clearly evident in Prince Rātburī's argument of January 1913, echoed in his preface to Ward's report two years later, that in the absence of a major irrigation initiative, in time Siam's rice would be unable to compete in international markets against rice from Burma and Cochin-China, and that as a result the kingdom would become increasingly impoverished. In short, the government's commitment to irrigation in the 1910s was considerably more securely based than it had been in the previous decade. That the commitment did not lead to the implementation of Ward's report in full can be ascribed to the inability of the administration to secure the required capital resources in Europe, as it had originally intended, following the outbreak of war in August 1914. It was this financial obstacle, not a weakness in the government's commitment, which determined the fate of Ward's irrigation proposals.

There remains the argument, advanced by Feeny, that opposition to major irrigation projects in the Central Plain from the landowners at Rangsit, many of them senior members of the Bangkok administrative élite, had a powerful influence in the decision-making of the government. In effect, the argument indicates that the private interests of the administrative élite were allowed to override the wider interests of the population as a whole. Feeny's analysis in this respect draws primarily on a brief passage from a Royal Irrigation Department report published in 1927, and concerned with the administration's decision in 1915 to give preference to the Pāsak project rather than to the Suphan scheme: 'The Government took the above course probably because it was considered inadvisable to disturb existing arrangements of landlords and tenants in the Rangsit area and elsewhere, which the opening up of big areas of land in Subhan, free for all, must have done.'[111] Feeny's argument must be treated with caution. First, it may be questioned whether evidence which relates specifically to the Siamese administration's preference for one relatively modest irrigation scheme over another in 1915 may be used to draw conclusions with respect to the government's irrigation policy as a whole over the period from the early 1900s. Thus, even if it is accepted that the reason for the administration's rejection (but in effect, simply postponement), of the Suphan scheme in 1915 was the desire to protect the private interests of the Bangkok élite in Rangsit, it does not necessarily follow that that same consideration was of importance in the government's rejection of van der Heide's major irrigation proposals in the previous decade.

Furthermore, it is possible to argue that Feeny has drawn from the passage cited an inference more precise than in fact it will bear. Although the statement that the government 'considered [it] inadvisable to disturb existing arrangements of landlords and tenants in the Rangsit area and elsewhere', may well be an oblique reference to a wish on the part of the administration not to inflict financial losses on the Rangsit landowners through the development of major new irrigated districts, it is clear that other inferences may be drawn. For example, the government may have wished to avoid the disruption of provincial administration and the serious disturbance to rural economic and social structures which a large-scale migration of cultivating families from Rangsit, the premier rice-producing district in the kingdom, might have occasioned. Perhaps more importantly, as the Irrigation Department had assumed responsibility for the Rangsit district in 1914 on the expiration of the concession of the Siam Land, Canals, and Irrigation Company, and was now intent on restoring the canals and dikes which had seriously deteriorated under the latter's administration,[112] the government clearly would not have wished to disturb the existing pattern of settlement in that area by opening up new tracts which may have drawn off the Rangsit cultivators in large numbers.[113] In summary, whilst the passage cited by Feeny may indeed be said to indicate that in 1915 the administration was concerned not to provoke a large-scale migration from Rangsit, it offers no guidance as to why that migration would have been so unwelcome to the authorities.[114] The threat posed to the private interests of the Rangsit landowners may, or may not, have been the fundamental concern. On this crucial point there can be merely conjecture.

Two further criticisms may be made of Feeny's analysis. First, it is difficult to see why the Bangkok administrative élite, having used its political and economic power to secure large landholdings in Rangsit at the end of the nineteenth century, would not have welcomed the opportunities provided by the principal van der Heide and Ward proposals to expand its landownership into further prime rice-cultivating districts and thereby strengthen its *rentier* position. Indeed as the condition of the Rangsit canals, dikes, and embankments seriously deteriorated under the administration of the company, a deterioration which presumably threatened to depress land rents in the area,[115] there would have been a particularly powerful incentive for the major élite landowners to relocate their investments in newly-opened tracts where truly effective irrigation facilities could justify premium rents.[116]

Finally, it is difficult to accept without reservation the inference which Feeny draws from the observation that among the Rangsit landlords were powerful members of the Bangkok administrative élite. The validity of the observation itself is beyond dispute. The

Rangsit landowners included, most notably, members of the Sanitwongse family (Phra-ongčhao Sai Sanitwongse, Chulalongkorn's personal physician, was a founder partner of the Siam Land, Canals, and Irrigation Company; Sai's eldest son, Suvabhan, was prominently involved in the company; his second son, Čhaophrayā Wongsānupraphat, was Minister of Agriculture in the years 1909–12);[117] the King's half-brothers, Prince Phichit Prīchākǫn and Prince Narāthip Praphanphong (Minister of Finance in the early 1890s);[118] and, according to one source, Chulalongkorn himself.[119] Neither may it be seriously doubted that the financial interests of the élite landowners in Rangsit would have suffered if new areas of the Central Plain had been opened up by major irrigation works. However, these observations, in themselves, do not constitute firm evidence that the élite landowning interests were a dominant influence when the Siamese administration came to consider the various proposals for large-scale irrigation works placed before it in the 1900s and 1910s. It is thus important to note that Feeny's argument for the influence of élite interests receives no support from the documentary record of the administration's deliberations on irrigation in this period. His evidence is simply circumstantial. Given this context, and given also the other influences which it is suggested shaped the government's irrigation policy in the early twentieth century, but influences for which there *is* firm and direct evidence, it is difficult to accept that private élite interests had the force proposed by Feeny. If the Siamese Government had been able to secure sufficient capital resources to implement a major irrigation programme in the Central Plain in this period, can it be seriously anticipated that that programme would have been blocked by the influence of élite interests?[120] Can the self-interest of certain members of the Siamese administrative élite as landowners in Rangsit be accepted as a more powerful determinant of the government's irrigation policy in the opening two decades of the twentieth century than the cost of a major irrigation initiative, in relation both to the volume of resources at the government's command and to the demands of the other major expenditure programmes before the administration?

VI

It is important to set the irrigation administration of the Siamese Government in the early twentieth century against that of the neighbouring colonial regimes in Lower Burma and Cochin-China in the same period, not least because it is a comparison which at times significantly influenced the Siamese Government itself.[121] Unfortunately, a number of difficulties are encountered in developing such a comparison. For example, it is virtually impossible to calculate the scale of the British and French administrations' investments undertaken specifically to improve irrigation and drainage for the cultivation of rice, for the majority of large earth-moving projects in the Irrawaddy and Mekong deltas not only influenced the water conditions for agriculture but were also primarily intended to improve inland water communications. More importantly, there are sufficient differences in climatic and topographical features between the Irrawaddy, Čhaophrayā, and Mekong deltas to demand water-control works of markedly different scale and form in the three deltas if the cultivation of rice as a major export commodity was to be ensured. Thus, for example, in Lower Burma the rainfall is fully sufficient for the cultivation of this crop; indeed the problem of water control here lies in protecting the fields against serious flooding when the Irrawaddy is in spate.[122] In contrast, the average rainfall in the Central Plain of Siam during the growing season provides less than two-thirds of the amount of water ideally required for the cultivation of paddy, and consequently, in the absence of irrigation works, successful cultivation depends on the Čhaophrayā River rising to a level sufficient to flood.[123] With respect to the Mekong delta, the average rainfall is substantially above that in the delta of the Čhaophrayā;[124] but the rains are also very irregular,[125] and this undermines considerably the security of cultivation. One distinctive feature of the topography of the Mekong delta should also be noted. The elevation of the Cochin-China plain seldom reaches the height of the South China Sea's highest tides, and consequently substantial areas of the delta have notably poor drainage and are prone to heavy inundation of sea water.[126] In these conditions the permanent settlement and cultivation of the unoccupied districts of the delta, specifically the trans-Bassac region, demanded the excavation of a major network of canals to drain the land and to secure communication into barely-inhabited areas.

Having established that differences in climatic and topographical conditions between the Irrawaddy, Čhaophrayā, and Mekong deltas demanded water-control works of markedly different scale and form, it can now be argued that throughout the colonial period both the British administration in Lower Burma and the French administration in Cochin-China undertook water-control projects on the minimum scale necessary to secure the cultivation of rice as a major export commodity. In Lower Burma, this involved the colonial administration in the construction of a series of great embankments along the banks of the Irrawaddy and its principal distributaries, notably in the Henzada district, to reduce the incidence of flood damage.[127] In Cochin-China this involved the French in a major feat of engineering—the excavation of an extensive complex of communication/drainage canals that would bring large tracts into cultivation.[128] However, neither the British nor the French colonial administration was prepared to undertake the construction of the large-scale irrigation and drainage works which would have enabled the rice farmers of the Irrawaddy and Mekong deltas to draw the full productive potential from the lands they cultivated. This crucial point is readily acknowledged in the secondary literature, even by authors who are broadly sympathetic to the colonial administrations. Thus B. O. Binns, a senior agricultural official in pre-war Burma, wrote in 1943:

> Major drainage works have been discussed, notably in respect of the drainage of the Pegu Plain, but so far as I know no major drainage work, as such, has been carried out, such drainage works as have been carried out being incidental effects of works intended mainly to achieve other purposes.... The existing irrigation works run by the [Irrigation] Department are entirely confined to Upper Burma.... No attempt has been made by Government to provide irrigation facilities in the precarious areas of Lower Burma where so much depends on the October rainfall.[129]

Charles Robequain, writing in 1939 of the provision of irrigation and drainage facilities in Cochin-China, was equally critical:

> ... the layout of the large canals, especially the older ones, was not adapted to the specific needs of rice-growing ... water level changes as yet benefit only a small number of rice plantations located along the main canals. The construction of secondary and tertiary canals lags well behind that of the principal ones, which is like a circulatory system lacking the smaller arteries and capillaries. Most of the rice plantations in Cochin China are not really irrigated. The upper part of the delta, around Chau Doc and Long Xuyen, is exposed to the floods of the Mekong which are often severe and necessitate the cultivation of wet rice with a mediocre yield.[130]

The evidence from Binns and Robequain suggests that Prince Rātburī's assessment in 1913 of the irrigation achievements of the neighbouring British and French administrations was unjustifiably generous.[131] In relation to the varied water-control requirements of the Irrawaddy, Čhaophrayā, and Mekong deltas, the achievements of the colonial regimes throughout the decades of Western rule do not appear markedly superior to that of the Siamese administration in the period under study.

Without a detailed examination of the colonial records comparable to that of the Siamese material which has been laid out in this chapter, it is perhaps unwise to venture an explanation for the failure of the British and French to undertake a decisive investment in irrigation. Nevertheless, a number of observations on this aspect of colonial administration in Lower Burma and Cochin-China can be drawn from the relevant secondary sources, and these observations in turn provide further insight into the irrigation administration of the Siamese Government in this period.

Undoubtedly, the extremely limited provision of effective irrigation in the Irrawaddy and Mekong deltas in this period can be explained in part by circumstances which were not only distinctive to those territories but which were also outside the immediate influence of the colonial authorities. For example, the crucial failure to construct the complex of secondary and tertiary canals in the French colony was explained by Robequain in terms of 'the feverish speculation which [has] characterized the development of eastern Cochin China for fifty years' and the landowners' 'haste to get the land into production'.[132] Later writers have emphasized either the unimaginative, insular self-interest of the principal Cochin-Chinese landowners, or the nature of landlord/tenant production relations in the delta. Thus, according to Robert L. Sansom:

.:. the landed oligarchy was static in attitude and motivation. It seemed to combine the aversion to technology of the pre-French mandarins with the isolation of the French bureaucrat or administrator. Living in Saigon or the larger provincial cities, the landlords had little knowledge of the problems of the peasants on their land and a negligible desire to solve them. Their only concern was to collect rents.[133]

And Martin J. Murray states:

The landlord/tenant production relations [in the Mekong delta] provided neither the tenants nor the landlords with inherent incentives to rotate crops and to take steps to avoid soil deterioration, to make seed selections on the basis of long-term yields, and to control the vagaries of the natural environment through extensive improvements in dike maintenance and construction.[134]

But in considering the provision not of subsidiary channels, but of the large-scale irrigation works whose construction could be undertaken only by the state, it is clear from the secondary literature that the perceptions and achievements of the colonial administrations in this respect were in large part shaped by precisely those considerations which were found to be most potent in Siam—that is, tight restriction on the financial resources of the state, and an, arguably enforced, preference on the part of the authorities for expenditure projects which would secure administrative or strategic advantages. Thus Robequain notes that the French administration undertook to finance public works very largely from its ordinary budget, for the scale of Indo-China's public debt was limited by the difficulty of increasing the tax burden of a population whose living standards remained low.[135] In Burma, capital restriction arose from the financial structure of the British India administration, under which a very substantial proportion of the revenues raised in the province was appropriated by the Central Government of India but which in turn maintained only the most modest allocation for the province's capital projects, including the construction of railways and of irrigation works.[136] 'The financial history of Burma ... is a record of persistent applications by the Local Government for an enhanced share of revenue for development purposes, and as persistent refusals by Imperial Government,' wrote a former Lieutenant-Governor of Burma, Sir Harvey Adamson, in 1918.[137] The province's revenues thus appropriated were used to secure the wider interests of British India, notably the relief of famine, the defence of the North-West Frontier, and the construction of various railways in the subcontinent.[138] In Indo-China, well over half the government's expenditure on economic infrastructure in the period 1900–35, in itself the single largest item of the colonial budgets, was devoted to the construction of railways and roads.[139] In detail, from 1918 the French administration embarked on a major programme of road building into all parts of its Indo-China possessions, so that, according to Robequain, by the 1930s the territory had one of the finest road systems in the Far East.[140]

The evidence above suggests that in colonial Burma and Indo-China, as in Siam, the need to secure external defence and internal order, and the need to facilitate effective administration from the main centre(s) of government, imposed demands on the state's financial resources which could not be denied. In contrast, even without a decisive irrigation initiative, the Irrawaddy and Mekong deltas, along with the delta of the Čhaophrayā, would still emerge as the most productive rice-exporting regions in the world.[141] The evolution of this common experience raises a central theme of this study, to which I will return.

1. Shigeharu Tanabe, 'Land Reclamation in the Chao Phraya Delta', in Yoneo Ishii (ed.), *Thailand: A Rice-Growing Society*, Honolulu, 1978, p. 45. The opening section of this chapter (pp. 8–11) draws on this article as a whole, as well as on Shigeharu Tanabe, 'Historical Geography of the Canal System in the Chao Phraya River Delta, from the Ayutthaya Period to the Fourth Reign of the Ratanakosin Dynasty', *Journal of the Siam Society*, vol. 65, pt. 2 (July 1977), pp. 23–72, and David B. Johnston, 'Rural Society and the Rice Economy in Thailand, 1880–1930', Ph.D. diss., Yale University, 1975, pp. 42–52.

2. This observation has important analytical implications which must be briefly noted. Recent research by Thai scholars into the earlier economic and political structures of Siam and their modulations in the more modern period, research which will be considered in the concluding chapter of this study, has drawn heavily on the concept of the 'hydraulic state' as advanced by Karl A. Wittfogel (*Oriental Despotism: A Comparative Study of Total Power*, New Haven, 1957), and in particular on its intellectual precursor, the concept of the Asiatic Mode of Production as advanced by Marx. Central to these concepts is the view that the Asiatic state formed an essential part of the productive base of society, for the state alone had the organizational capacity to construct and superintend the hydraulic works without which large-scale agricultural production in those societies could not be maintained. Tanabe's analysis, that the pre-modern Siamese state tended 'to emphasize communication rather than agricultural production in the large-scale hydraulic works' ('Land Reclamation in the Chao Phraya Delta', p. 51), thus implicitly challenges the applicability of an important part of the Wittfogel and Marx concepts to the Siamese historical experience.

3. Sir John Bowring himself observed, in an oft-quoted comment, that the 1855 Treaty 'involved a total revolution in all the financial machinery of the Government' (Sir John Bowring, *The Kingdom and People of Siam*, reprinted Kuala Lumpur, 1969, vol. II, p. 226). In particular the Treaty required the abolition of all forms of royal trading monopoly. See also Tanabe, 'Historical Geography of the Canal System in the Chao Phraya River Delta', p. 55, note 82.

4. In fact, according to Tanabe ('Historical Geography of the Canal System in the Chao Phraya River Delta', p. 63), Phra Phāsī Sombatbǫribūn appears ultimately to have relied upon revenue from a contract for opium-tax farming to finance excavation. Johnston, op. cit., pp. 44–5, emphasizing the confusion of the sources on this point, notes that according to Chulalongkorn, although Phra Phāsī Sombatbǫribūn had proposed the construction of the Phāsī Čharoen canal and had sought tax-collecting concessions to undertake the project, the canal was in fact excavated by the government.

5. Johnston, op. cit., pp. 51–2.
6. This discussion of the Siam Land, Canals, and Irrigation Company in the period to the early 1900s (pp. 11–13) draws on Johnston, op. cit., pp. 52–91.
7. Sommot to Rivett-Carnac, 15 December 1899; Chulalongkorn to Rolin-Jacquemyns, 15 December 1899, NA r5 KS 9.3/9.
8. Rivett-Carnac to Čhaophrayā Thēwētwongwiwat, 6 December 1899, NA r5 KS 9.3/9.
9. Čhaophrayā Thēwēt to Chulalongkorn, 23 November 1899, NA r5 KS 9.2/18.
10. Čhaophrayā Thēwēt, 'Rayathāngtruat thīkhutkhlǭng' ('An Inspection Tour of Canal Excavations'), 15 June 1900; Čhaophrayā Thēwēt to Chulalongkorn, 1 January 1901, NA r5 KS 9.2/28.
11. Čhaophrayā Thēwēt to Chulalongkorn, 23 November 1899, NA r5 KS 9.2/18; Čhaophrayā Wongsānupraphat, *Prawat krasuangkasētrāthikān* (*History of the Ministry of Agriculture*), Bangkok, 1941, p. 155.
12. Chulalongkorn to Chaophrayā Thēwēt, 26 November 1899, NA r5 KS 9.2/18.
13. Rolin-Jacquemyns to Chulalongkorn, 25 December 1899, NA r5 KS 9.3/9.
14. Documents in NA KS(Ag) 11/1. Van der Heide was first approached in early 1902.
15. Ibid.
16. Monthly reports from van der Heide to Čhaophrayā Thēwēt, June–September 1902, NA r5 KS 9.2/39.
17. Tej Bunnag, 'Khabotngiao mu'angphrāe r.s. 121' ('The 1902 Shan Rebellion at Phrāe'), *Sangkhomsāt parithat*, vol. 6, no. 2, (September–November 1968), pp. 67–80; David F. Holm, 'The Role of the State Railways in Thai History, 1892–1932', Ph.D. diss., Yale University, 1977, pp. 113–15.
18. Ian Brown, 'The Ministry of Finance and the Early Development of Modern Financial Administration in Siam, 1885–1910', Ph.D. diss., University of London, 1975, pp. 119–49; Holm, 'The Role of the State Railways in Thai History', pp. 115, 121–6.
19. In one respect, van der Heide overstated the degree of water insufficiency in the Central Plain.

Water Conditions in the Central Plain, 1831–1948

Condition	Number of Years	Percentage
High Flood	4	3
Sufficient Water to Mature the Crop	53	45
Moderate Drought	21	18
Severe Drought	35	30
Extreme Drought	4	3
	117	

Source: David Feeny, *The Political Economy of Productivity: Thai Agricultural Development, 1880–1975*, Vancouver and London, 1982, p. 60.

This table suggests that sufficient rainfall was not in fact an 'extremely rare' occurrence in the Central Plain, for in 53 of the 117 years covered, water conditions were sufficient to mature the rice crop. However, it must be added that the Central

Plain did experience drought conditions in no less than 60 years in this period, and that in one year in three the drought was either severe or extreme.

This is an appropriate point to note also that the pattern of flooding and drainage varies quite markedly within the delta. The Čhaophrayā River, carrying with it vast quantities of silt, is continuously building up, and then breaking through, natural levees, with the result that the lower Central Plain, far from being flat, undulates with differences in height of several metres. Consequently within the delta there are upland areas in which rice cultivation is possible only if there is relatively severe flooding; lower areas which are flooded every year and which thus constitute the main districts within the delta in which swamp rice is cultivated; and low-lying areas which become deeply submerged after only moderate rainfall. See, Wolf Donner, *The Five Faces of Thailand: An Economic Geography*, London, 1978, pp. 232–3, 236–7, 764.

20. A number of sources (including Čhaophrayā Wongsānupraphat, op. cit., p. 156) suggest that the estimated cost of van der Heide's proposals was approximately 60 million baht. At the current rate of exchange, this was the equivalent of some £3,150,000; 47 million baht was equivalent to some £2,500,000.

21. Prince Devawongse to Čhaophrayā Thēwēt, 10 May 1903, NA KS(Ag) 11/71.

22. Čhaophrayā Thēwēt to Chulalongkorn, 9 May 1903, NA r5 KS 9/5.

23. Čhaophrayā Thēwēt to Chulalongkorn, 15 August 1903, NA r5 KS 9/4.

24. Prince Devawongse to Chulalongkorn, 24 August 1903, NA r5 Kh 5.1/19; Čhaophrayā Thēwēt to Chulalongkorn, 12 October 1903, NA r5 KS 9/4.

25. James C. Ingram (*Economic Change in Thailand 1850–1970*, Stanford, 1971, p. 197) states that van der Heide's full scheme was in fact rejected at this time. He appears to base his statement on a memorandum by Rivett-Carnac which argued very strongly against the irrigation proposals. ('Memorandum by the Financial Adviser upon the Cash Balances of the Government', 25 November 1903, NA K Kh 0301.1.30/6. Ingram erroneously attributes the memorandum to Rivett-Carnac's successor, W. J. F. Williamson.) However, the Financial Adviser's memorandum is dated late November 1903, whilst according to Čhaophrayā Thēwēt (letter to Chulalongkorn, 12 October 1903, NA r5 KS 9/4), the committee had concluded its discussions with respect to irrigation at least six weeks earlier.

26. Čhaophrayā Wongsānupraphat, op. cit., pp. 157–60.

27. 'Monthly Report of the Royal Irrigation Department', September 1904, NA r5 KS 9/7.

28. Van der Heide, 'Note in regard to the Cost of the Irrigation Scheme at Reduced Capacity, as far as to be Executed in the years 125/128 Inclusive, in connection with the City Water Works', 12 March 1906, NA KS(Ag) 11/87; NA K Kh 0301.1.18/5.

29. In fact 24 million baht spread over four years represented a higher average *annual* outlay on irrigation than 47 million baht extended over twelve years, as van der Heide had proposed in 1903.

30. Van der Heide to Prince Čhanthaburī, 18 February 1908, NA K Kh 0301.1.18/5. The Bangkok water supply scheme, and its relationship with van der Heide's irrigation proposals, will be considered briefly below.

31. Van der Heide to Prince Čhanthaburī, 18 February 1908, NA K Kh 0301.1.18/5; van der Heide to Čhaophrayā Thēwēt, 20 May 1907, NA r5 KS 1/11; Leslie E. Small, 'Historical Development of the Greater Chao Phya Water Control

Project: An Economic Perspective', *Journal of the Siam Society*, vol. 61, pt. 1 (January 1973), p. 3.

32. W. J. F. Williamson, 'Further Proposed Schemes of Irrigation Department', 14 August 1908; van der Heide, 'Note in Reference to the Pasak Scheme', 2 July 1908, NA K Kh 0301.1.18/4.

33. W. J. F. Williamson, 'Further Proposed Schemes of Irrigation Department', 14 August 1908, NA K Kh 0301.1.18/4.

34. Mọm Anuruthathēwā, 'Rāingān kromkhlǭng klāwthu'ng tangtāe tangkromkhlǭng čhonthu'ngtang Mọm Anuruthathēwā penčhaokrom' ('Report on the Canal Department from its Establishment to the Appointment of Mọm Anuruthathēwā as Director'), NA KS(Ag) 11/803.

35. Čhaophrayā Wongsānupraphat, op. cit., p. 156.

36. The Siamese administration was quite prepared to dispense with the services of its senior European employees if there was serious dissatisfaction with their performance. For example, in 1898 the government indicated firmly to the first Financial Adviser, Alfred Mitchell-Innes, that it would not renew his initial contract when it expired the following year, after complaints from, among others, the Minister of Finance, as to the poor quality of his work and his abrasive, insensitive, character. See, Brown, 'The Ministry of Finance', pp. 105–12.

37. This is an appropriate point to note that Mọm Anuruthathēwā (M. R. W. Saiyut Sanitwongse) was a member of the Sanitwongse family. As noted earlier, Phra-ongčhao Sai Sanitwongse was a founder partner of the Siam Land, Canals, and Irrigation Company, and the family maintained very substantial land investments in the Rangsit area. A number of scholars have argued that the implementation of van der Heide's scheme would almost certainly have threatened the financial interests of the Rangsit landlords (pp. 34–5). These circumstances may have been an additional motive for Mọm Anuruthathēwā's complaints against van der Heide, although the speculative nature of this suggestion requires no emphasis.

38. Documents in NA r6 KS 4/1.

39. In this context it might be added that because of serious design and construction failures, considerable damage was caused to the Rangsit locks in 1908. Following the very heavy monsoon of that year, the Canal Department ordered the opening of the locks in Rangsit, for it feared that there would be serious crop losses if the water level in that area was not lowered quickly. However the locks had not been built for the release of water, and were thus severely damaged during this operation. (See, Small, op. cit., p. 21.) As noted earlier, in 1910 Mọm Anuruthathēwā argued that the locks on the Phāsī Čharoen and Damnoen Saduak canals similarly contained no facilities for the release of excess water, a defect which he attributed to van der Heide's incompetence. At the very least, the damage inflicted on the Rangsit locks in 1908 and then the serious flooding from the Phāsī Čharoen and Damnoen Saduak canals in 1909 would have done little for the professional reputation of the Director of the Canal Department.

40. L. R. de la Mahotière to Čhaophrayā Thēwēt, 15 June 1903, NA K Kh 0301.1.18/2.

41. In answering de la Mahotière's further argument that the irrigation scheme would have a deleterious impact on the proposed Bangkok water supply system, van der Heide stated: 'The objections ... would be funny, if they were not founded on misunderstanding of the hydrotechnical side of the irrigation system.' And again: 'Such statements as these of Mr. L. R. de la Mahotière do not sound much like

expert argument.' Van der Heide to Čhaophrayā Thēwēt, July 1903, NA K Kh 0301.1.18/2; NA KS(Ag) 11/87.

42. Documents in NA r5 KS 9/8 and 9/11.
43. Čhaophrayā Thēwēt to Chulalongkorn, 12 October 1903, NA r5 KS 9/4.
44. *Bangkok Times*, 27 February 1913.
45. Van der Heide to Čhaophrayā Thēwēt, 20 May 1907, NA r5 KS 1/11.
46. Comment by King attached to Phrayā Srīsunthǫnwōhān to Chulalongkorn, 3 May 1909, NA r5 KS 9/11.
47. Prince Čhanthaburī to Chulalongkorn, 17 March 1909, NA r5 Kh 5.1/27. A further possible influence on the government's decision in 1909 is suggested by Small, op. cit., p. 3: 'In 1908 the worst flood in 30 years occurred, causing serious damage to many of the canal control structures which had been constructed by the Department of Canals in the years since 1903. Although there is no record of the effect of these events on the attitudes of the government ministers and advisers, it seems probable that they strengthened the position of those who opposed the irrigation proposals.' See also note 39.
48. Van der Heide to Mǫm Chātadētudom (Deputy Minister of Agriculture), 3 and 19 April 1909, NA K Kh 0301.1.18/5.
49. Phrayā Srīsunthǫnwōhān to Chulalongkorn, 3 May 1909, NA r5 KS 9/11.
50. Ibid.; *Bangkok Times*, 1 May 1909.
51. Phrayā Srīsunthǫnwōhān to Chulalongkorn, 3 May 1909, NA r5 KS 9/11.
52. *Bangkok Times*, 1 May 1909, 13 June 1909. With van der Heide's despondent departure from Siam, this is perhaps an appropriate point to note that he nevertheless enjoys a fine reputation among radical Thai scholars of the present day, including notably Chatthip Nartsupha and Suthy Prasartset. His reputation in this respect rests on an article he published during his time in the kingdom, 'The Economical Development of Siam during the Last Half Century', *Journal of the Siam Society*, vol. 3, pt. 2 (1906), pp. 74–101. This work has been reprinted as the opening document in Chatthip Nartsupha and Suthy Prasartset (eds.), *The Political Economy of Siam 1851–1910*, Bangkok, 1981, pp. 73–112, the editors regarding it as 'the first systematic analysis of the evolution of the political economy of Siam' (p. 71).
53. *Report of the Financial Adviser on the Budget of the Kingdom of Siam*, issues for the years 1910–11 to 1913–14.
54. Ministry of Agriculture, *Prawat krasuangkasēt (History of the Ministry of Agriculture)*, Bangkok, 1957, pp. 108–11.
55. See, for example, Small, op. cit., pp. 3–4; Ingram, op. cit., p. 83.
56. Ministry of Agriculture, op. cit., p. 124.
57. Prince Rātburī to Vajiravudh, 13 January 1913, NA r6 KS 4/2.
58. Ibid.
59. Meeting of the Council of Ministers, 20 January 1913; Vajiravudh to Prince Rātburī, 20 February 1913, NA r6 KS 4/2.
60. *Bangkok Times*, 23 June 1913; T. R. J. Ward, *Report on a Scheme for the Irrigation of so much of the Valley of the Menam Chao Bhraya as may be Possible for a Capital Outlay of One and Three Quarter Millions Sterling*, Bangkok, 1915, pp. ii–iii.
61. Ministry of Agriculture, op. cit., pp. 126–7; Small, op. cit., p. 4.
62. Ward, op. cit.
63. Ibid., p. iii.
64. Ibid., pp. 3–4.

65. Prince Čhanthaburī to Vajiravudh, 20 February 1915, NA r6 Kh 1/29.
66. Vajiravudh to Prince Čhanthaburī, 23 February 1915, NA r6 Kh 1/29.
67. Prince Čhanthaburī to Vajiravudh, 13 March 1915, NA r6 Kh 1/29.
68. Expenditure for construction of the northern railway was maintained at well over one million baht per annum in the years from 1915, the funds being drawn from the Treasury reserve. *Report of the Financial Adviser on the Budget of the Kingdom of Siam*, issues for the years 1917–18 to 1920–1.
69. Prince Rātburī to Vajiravudh, 28 April 1915, NA r6 KS 4/2.
70. Vajiravudh to Prince Rātburī, 3 May 1915, NA r6 KS 4/2.
71. Prince Čhanthaburī to Vajiravudh, 16 September 1915, NA r6 KS 4/2.
72. These calculations appear to have remained a matter of major dispute within the administration for some time. Small, op. cit., p. 15.
73. Quoted in, Small, op. cit., p. 5.
74. Ibid.
75. Ministry of Agriculture, op. cit., p. 128.
76. *Report of the Financial Adviser on the Budget of the Kingdom of Siam*, for the year 1920–1.
77. Ministry of Agriculture, op. cit., pp. 128–9; Irrigation Department, 'Report Regarding Revised Programme of Expenditure for the Pasak Project', 1 October 1918, NA KS(Ag) 11/1103.
78. Ministry of Agriculture, op. cit., p. 139; Small, op. cit., pp. 5–6.
79. An enlarged edition, *Economic Change in Thailand 1850–1970*, was published in 1971. References here are to the later edition.
80. Ingram, op. cit., pp. 196–202.
81. Ibid., p. 200.
82. Feeny, *The Political Economy of Productivity*, pp. 81–2.
83. Ibid., pp. 80–4.
84. Ibid., p. 84.
85. Small, op. cit., pp. 8–9, 14–17.
86. Johnston, op. cit., pp. 92–4.
87. For a general consideration of the influence of the British Financial Advisers in this period see, Ian Brown, 'British Financial Advisers in Siam in the Reign of King Chulalongkorn', *Modern Asian Studies*, vol. 12, pt. 2 (April 1978), pp. 193–215.
88. In British India, cultivators invariably paid for irrigation water from government works, usually through a specific irrigation rate which was varied according to the acreage and nature of the crop. See, for example, Nasim Ansari, *Economics of Irrigation Rates: A Study in Punjab and Uttar Pradesh*, London, 1968, particularly chapter 2.
89. Čhaophrayā Thēwēt to Chulalongkorn, 23 November 1899, NA r5 KS 9.2/18.
90. Ingram, op. cit., p. 199, note 11; Small, op. cit., p. 15.
91. Ingram, op. cit., p. 199, note 11.
92. See p. 10.
93. Ingram, op. cit., p. 82.
94. Small, op. cit., p. 15.
95. Prince Rātburī to Vajiravudh, 28 April 1915; Prince Čhanthaburī to Vajiravudh, 16 September 1915, NA r6 KS 4/2.
96. Feeny, *The Political Economy of Productivity*, p. 83.
97. Brown, 'The Ministry of Finance', pp. 153–6.

98. Prince Rātburī to Vajiravudh, 13 January 1913, NA r6 KS 4/2.
99. Ward, op. cit., p. 4. Emphasis added.
100. Prince Čhanthaburī to Vajiravudh, 20 February 1915, NA r6 Kh 1/29.
101. Ingram, op. cit., pp. 181-2. The direct financial return to the government from the employment of its loan funds in this way would have been limited to the interest earned on that portion of the reserves held on deposit in Europe.
102. The view that van der Heide's main proposals were rejected because of their high cost was advanced by a number of contemporary sources, including: Prince Devawongse to Čhaophrayā Thēwēt, 10 May 1903, NA KS(Ag) 11/71; Report by Mǫm Anuruthathēwā, NA KS(Ag) 11/803; Ministry of Agriculture, op. cit., pp. 69-70; Čhaophrayā Wongsānupraphat, op. cit., p. 156; *Bangkok Times*, 27 February 1913.
103. Again it should be noted that a number of sources indicate that the estimated cost of van der Heide's proposals was approximately 60 million baht. See note 20.
104. *Report of the Financial Adviser on the Budget of the Kingdom of Siam*, for the year 1905-6.
105. See, for example, Ingram, op. cit., pp. 198, 200; Feeny, *The Political Economy of Productivity*, p. 84.
106. Anderson has argued that the strengthening of the Siamese armed forces from the reign of Mongkut was undertaken not to reinforce the external security of the kingdom, for the advance of European rule in mainland South-East Asia had demilitarized Siam's traditional rivals, but rather was 'mainly a means for *internal* royalist consolidation'. See Benedict R. O'G. Anderson, 'Studies of the Thai State: The State of Thai Studies', in Eliezer B. Ayal (ed.), *The Study of Thailand: Analyses of Knowledge, Approaches, and Prospects in Anthropology, Art History, Economics, History, and Political Science*, Athens, Ohio, 1978, pp. 200-5. For this argument Anderson drew on Noel Alfred Battye, 'The Military, Government and Society in Siam, 1868-1910: Politics and Military Reform during the Reign of King Chulalongkorn', Ph.D. diss., Cornell University, 1974. However, it might be added that with respect to the challenge which the European powers themselves posed to Siam's political independence, particularly from the last decade of the nineteenth century, the consolidation of Chakri internal power was in itself crucial in defending the external security of the kingdom. A serious disintegration of internal authority would almost certainly have provoked or enticed European armed intervention.
107. Yet the demands on the government's resources were so severe in this period that even construction of the northern railway could proceed only slowly. It was not until January 1922 that the final stretch to Chiangmai was officially opened. Holm, 'The Role of the State Railways', p. 186.
108. Indeed I have argued elsewhere that 'the volume of resources devoted to the exchange stabilization fund was no more than was required for the fund to discharge its function, at least in the late 1900s'. Ian Brown, 'Siam and the Gold Standard, 1902-1908', *Journal of Southeast Asian Studies*, vol. 10, no. 2 (September 1979), p. 399.
109. However, it should be added that in presenting his 1908 proposals, van der Heide explicitly stated that the projects envisaged, although self-contained, were designed as integral to the 'big scheme'. (W. J. F. Williamson, 'Further Proposed Schemes of Irrigation Department', 14 August 1908; van der Heide, 'Note in Reference to the Pasak Scheme', 2 July 1908, NA K Kh 0301.1.18/4.) The Siamese

ministers may therefore have seen the 1908 proposals as simply a stratagem by van der Heide to ensure the eventual sanction of his principal report of 1903.

110. Documents in NA KS(Ag) 11/1. Emphasis added.

111. Feeny, *The Political Economy of Productivity*, pp. 83-4. For a clear statement of the wider argument implicitly drawn from this passage, see ibid., p. 104; and David Feeny, 'Extensive versus Intensive Agricultural Development: Induced Public Investment in Southeast Asia, 1900-1940', *Journal of Economic History*, vol. 43, no. 3 (September 1983), p. 703.

112. Prince Čhanthaburī to Vajiravudh, 16 September 1915, NA r6 KS 4/2.

113. An official history of the Ministry of Agriculture (*Prawat krasuangkasēt*, p. 128) explained the preference of the government for the Pāsak scheme in 1915 directly and solely in terms of the Irrigation Department's recently assumed responsibility for the Rangsit district.

114. In a similar manner, W. J. F. Williamson's recommendation, as noted by Feeny (*The Political Economy of Productivity*, p. 82), 'that the Pasak project be pursued since it would benefit areas already under cultivation and would avoid the problem of migration out of the Rangsit area', cannot be said to constitute evidence of sufficient precision to sustain Feeny's argument.

115. Indeed Johnston implies that there was a serious decline in Rangsit land prices and rents in the final years of the company's concession (Johnston, op. cit., p. 60).

116. In apparent anticipation of this criticism, Feeny argues: '... ownership of most of the land in the Central Plain would have been difficult to arrange. By 1903 [the year in which van der Heide's main report was presented] élite who had speculated heavily in land at Rangsit were probably not in a financial position to attempt to extend their control over a much larger area' (*The Political Economy of Productivity*, p. 84). However, the argument above does not require the élite to have secured ownership of most of the land in the Central Plain, but simply a sufficient area to compensate them for possible losses in Rangsit. Nor, it should be added, is there evidence to sustain (or deny) the proposition that the élite had seriously weakened their financial position by their earlier speculation in the Rangsit district.

117. For a brief account of the Sanitwongse family and its various bureaucratic and business interests, see Johnston, op. cit., pp. 262-75.

118. Ibid., pp. 80-1.

119. Ammar Siamwalla, *Land, Labour and Capital in Three Rice-growing Deltas of Southeast Asia 1800-1940*, New Haven, July 1972, p. 28.

120. Cf., 'The van der Heide proposal and a number of other irrigation proposals were turned down because the élite were unable to capture the benefits.' Feeny, *The Political Economy of Productivity*, p. 104.

121. See, for example, Prince Rātburī to Vajiravudh, 13 January 1913, NA r6 KS 4/2.

122. Charles A. Fisher, *South-East Asia: A Social, Economic and Political Geography*, London, 1966, p. 436.

123. Ibid., p. 495.

124. Feeny, *The Political Economy of Productivity*, p. 186, note 1.

125. Charles Robequain, *The Economic Development of French Indo-China*, London, 1944, p. 222.

126. Ibid., pp. 221-2.

127. Fisher, op. cit., p. 437; J. S. Furnivall, *An Introduction to the Political*

Economy of Burma, Rangoon, 1931, p. 26.

128. One French writer placed the excavation of this canal network 'among the great works of modern civilization' (quoted in Robert L. Sansom, *The Economics of Insurgency in the Mekong Delta of Vietnam*, Cambridge, Mass., 1970, p. 48). Excavation began in the mid-1860s, largely utilizing manual labour, and it was only from the end of the century that increasingly powerful mechanized dredging equipment came to be used (Robequain, op. cit., pp. 110–11). By 1930 as much as 165 million cubic metres had been excavated in Cochin-China: this compares with an excavation of 260 million cubic metres in the construction of the Suez Canal, and 210 million in the excavation of the Panama Canal (Sansom, op. cit., p. 48).

129. B. O. Binns, *Agricultural Economy in Burma*, Rangoon, 1948, p. 63.

130. Robequain, op. cit., pp. 111–12, 221–2.

131. Prince Rātburī to Vajiravudh, 13 January 1913, NA r6 KS 4/2.

132. Robequain, op. cit., p. 111.

133. Sansom, op. cit., p. 52.

134. Martin J. Murray, *The Development of Capitalism in Colonial Indochina (1870–1940)*, Berkeley, 1980, p. 435.

135. Robequain, op. cit., pp. 156–7.

136. G. E. Harvey, *British Rule in Burma 1824–1942*, London, 1946, pp. 57–9; Maung Shein, *Burma's Transport and Foreign Trade, 1885–1914*, Rangoon, 1964, pp. 196–204.

137. Quoted in, Maung Shein, op. cit., p. 204.

138. Ibid., pp. 202–3.

139. Murray, op. cit., p. 174. For comparison, just 19 per cent of total colonial infrastructure expenditure was devoted to 'hydraulic works, irrigation projects, and navigation'.

140. Robequain, op. cit., pp. 98–9.

141. This is an appropriate point to add that the British and French were prepared to invest in substantial water-control projects in areas of threatened agrarian crisis (the dry zone of Central Burma and, in particular, the densely-populated Song-koi delta of Tonkin), where such works became essential for the maintenance of subsistence cultivation. Binns, op. cit., p. 63; Fisher, op. cit., p. 546.

2
The State and the Rice Economy: Technical Change and Economic Depression

IN the previous chapter it was held that the most effective way for the Siamese administration of the early twentieth century to have promoted an expansion in rice acreage and an increase in rice yields would have been through the construction of large-scale irrigation and drainage facilities in the lower Central Plain. Yet even with the failure of the government to commit itself to such a programme, there remained a number of further measures which also would have raised the productivity of the rice economy, although it might be added that the full benefit of many of these further measures would have been realized only if they were introduced in conjunction with major water-control works. The specific measures being considered here include the development and diffusion of high-yield rice seeds, the introduction of more advanced agricultural equipment, the encouragement of an increased use of fertilizer, and the evolution of more productive cycles of rice planting and harvesting. The principal concern of the first section of this chapter is to consider the Siamese administration's policies and performance in the application of these aspects of scientific agriculture to the rice economy of the early twentieth century.

I

If the principal constraint on the ability of the Siamese administration to undertake a major irrigation programme in the opening decades of the twentieth century was an acute restriction in capital resources, then the principal constraint on the implementation of a programme of scientific agriculture in the same period was an acute shortage of appropriately trained and experienced personnel. It was not simply that officials had to be found who could undertake research into the development of new varieties of rice seed and more productive agricultural practices, and into the effective use of fertilizers; it was also necessary to procure sufficient manpower to ensure the successful diffusion of these more advanced inputs and methods throughout the rice districts of Central Siam. Moreover, as the following pages will demonstrate, it was not sufficient for these officials to be trained solely in the techniques of scientific agriculture then being advanced in the Western world; it was essential that they should also have a firm practical knowledge of agricultural conditions and practices in Siam itself. Indeed it was this need for officials to have command of both Western techniques *and* local practices that was to prove the most intractable aspect of the agricultural administration's manpower crisis.

These problems first came to the fore towards the end of the 1890s. During that period, the establishment of a government experimental farm had frequently been discussed by the administration.[1] But nothing had come of these initiatives for, according to the King, the government did not possess the specialist knowledge that would enable it to proceed confidently with such a project, while it was also felt that to engage a foreign agricultural officer would be too expensive. However, a possible solution to these difficulties presented itself in June 1900 when the Minister of Education received a letter from the Siamese Minister in Paris informing him of the presence in France of a Siamese student, one Nāi Čharoen.[2] Nāi Čharoen, who came from a noble family, had been sent as a government scholar to France in 1892, when he was 17 years old. He had spent five years in acquiring a general education before moving to a college at Montargis, south of Paris, where he had spent a further three years studying agricultural science.[3] This last phase of his education had been quite limited. Nāi Čharoen had completed only the first of three levels of instruction and examination, and that instruction had been primarily theoretical rather than practical. Moreover, the education at Montargis had, of course, involved the study of agricultural science in the context of European conditions and practices—Nāi Čharoen knew little of the conditions of agriculture in his own country. Despite these shortcomings, the Minister of Education felt that on his return from France, Nāi Čharoen should be offered a probationary position in the Ministry of Agriculture.[4] The minister in that Department, Čhaophrayā Thēwēt, responded eagerly to this prospect, seeing in Nāi Čharoen the instrument by which the administration could at last initiate a programme of agricultural research.[5] But the King appears to have had some misgivings, for he noted in particular Nāi Čharoen's youth and lack of experience.[6]

Nevertheless, he was prepared to sanction his appointment to the Ministry of Agriculture, although he argued that his first task should be to undertake an inspection tour of agricultural districts near the capital and to prepare a report on his findings. Indeed the King laid out an itinerary for Nāi Čharoen.

It soon became clear that Chulalongkorn's misgivings were fully justified. An interview with the Minister of Education on his return from Europe revealed that Nāi Čharoen could converse comfortably only in French, and that his understanding in his mother tongue was distinctly patchy.[7] This language difficulty may explain in part the dismal manner in which Nāi Čharoen carried out his inspection tour of the central agricultural districts. His departure from Bangkok was at first delayed for more than four months by ill health, yet when he did eventually leave the capital he visited only one of his four stipulated provinces before scuttling back.[8] On his return, he submitted a sketchy three-page report which, according to the King, contained numerous factual errors.[9] Almost immediately, Nāi Čharoen was instructed by his Minister to complete his provincial tour. Although this appears to have been achieved without serious incident, according to Čhaophrayā Thēwēt, Nāi Čharoen's subsequent report indicated that he still had little understanding of the cultivation practices found in the kingdom.[10] Reporting to the King in August 1901, the Minister of Agriculture added that although Nāi Čharoen had been imploring him to proceed with the establishment of the experimental farm, when questioned in detail on this proposal the latter had been unable to respond to any effect. Indeed, the impression was given by Čhaophrayā Thēwēt that even mundane comments had to be dragged from a ponderous Nāi Čharoen. These issues came before the Council of Ministers in early September 1901.[11] Despite his misgivings, Čhaophrayā Thēwēt, supported by the Minister of Finance, Prince Mahit, argued for the early establishment of the experimental farm, but the King and Prince Damrong suggested that before any further action was taken Nāi Čharoen should be sent to Saigon to inspect the agricultural research being undertaken in the French colony. Unfortunately, his visit proved an almost complete disaster. In December 1901, the Siamese Consul in Saigon wrote to Prince Devawongse: 'As regards Mr. Nai Charon, full opportunity is afforded in order to enable him to carry out the work with which he is entrusted, but the general feeling amongst the official agricultural staff is that he seems not prepared to assume his charge, as it is most difficult to draw from him what kind of informations [sic] he requires.'[12] Chulalongkorn was furious at this loss of face before the French.[13] His anger was further fuelled when he read Nāi Čharoen's report on his four-month visit to French Indo-China, submitted on his return to Bangkok. Quite simply the King doubted whether Nāi Čharoen was its true author.[14] This suspicion brought his career in the agricultural administration to an apparent close.

In the present context, the case of Nāi Čharoen is noteworthy in two principal respects. First, attention must be drawn to the unplanned, improvised character of his agricultural education in France and his subsequent recruitment into the Ministry of Agriculture on his return to Siam. It is quite clear that Nāi Čharoen was not specifically selected and educated with a view to taking on specialist responsibilities of the kind assigned to him by the government from mid-1900.[15] Second, in view of the elementary nature of Nāi Čharoen's training in agricultural science and, perhaps more importantly, his apparently leaden manner, it is surprising that the administration should have seriously considered him as a potential mainspring for its proposed programme of agricultural research, and that it should have persevered with him for the length of time that it did. Almost certainly the government's over-optimistic and over-indulgent treatment of Nāi Čharoen arose primarily from the fact that it faced the most severe shortage of appropriately trained and experienced personnel in this specialist field. In difficult circumstances he appeared as the only option.

Nāi Čharoen's failure did not discourage continued demands for the establishment of an experimental farm. In early 1903, Kametaro Toyama, the head of a group of Japanese sericulture experts then attached to the Ministry of Agriculture,[16] undertook an inspection tour of Nakhǫn Chaisī to investigate rice cultivation in the province.[17] He observed that the Siamese farmers of that area concentrated so much on the cultivation of rice that they apparently knew little of the requirements for cultivating other profitable crops, and furthermore, that even in the cultivation of rice many farmers had little skill in selecting the most productive strains of seed. Toyama therefore called for a programme of cultivation experiments and trials. This call was vigorously taken up the following year by Prince Phenphatanaphong, a son of Chulalongkorn and, at the age of 22, Assistant Minister of Agriculture and Director of the Agricultural Department.[18] During that decade the

Prince was to emerge as perhaps the most perceptive and committed advocate within the administration of the advance of cultivation practices in the kingdom through governmental initiative and action.

Prince Phenphat first laid out his ideas in full on agricultural reform in a lengthy letter to his Minister, Čhaophrayā Thēwēt, in October 1904.[19] Predictably he urged government action. However, he also warned that to secure firm advances in cultivation methods in the kingdom would take a considerable time—many decades—and would involve frequent failures and set-backs. Specifically, Prince Phenphat explained that the administration would inevitably face considerable difficulties as it sought to create a body of distinctive agricultural expertise on which its programmes for the advance of cultivation practices could then draw. To establish this argument, he drew upon the contemporary experience of Japan. The Prince explained that in the 1870s, in pursuit of agricultural advance, the Japanese administration had established special agriculture schools and had employed instructors from Europe to teach in them. But when the Japanese had applied the agricultural practices which the Europeans had taught them, they found that the yields were lower than those achieved with established Japanese practices. Climatic and soil conditions in Japan differed so markedly from those in Europe that the direct application of advanced Western cultivation methods secured no advantage; indeed it was counter-productive. The agricultural schools were closed. According to Prince Phenphat, the Japanese had then adopted an alternative approach. In this, a number of Japanese with an intimate knowledge of local cultivation practices had been sent to Europe to study agricultural science, with the intention that they would compare methods and select those which could be applied in their own country. However, even this approach had failed, for most of the Japanese thus sent to Europe were simple farmers who did not have the educational background necessary to study and absorb the technical literature in agricultural science placed before them. This failure provoked a third approach. Selected Japanese students who had already acquired a sound general education were sent to Europe for three years to study agricultural science. On their return to Japan, they had undertaken the study of local cultivation practices. Thus armed with a detailed knowledge of both European scientific agriculture and Japanese agricultural conditions, the students had been well placed to promote major advances in indigenous agriculture. According to Prince Phenphat, this last approach—in creating a body of distinctive agricultural expertise on which programmes for the reform of cultivation practices had then drawn—had been notably successful. A range of textbooks on advanced cultivation methods had been produced; the agriculture schools had been reopened, now teaching cultivation practices specifically developed for Japanese conditions; and experimental farms had been established. In making these opening observations, Prince Phenphat was undoubtedly influenced by two considerations in particular—the administration's recent unfortunate experience with Nāi Čharoen, and the presence in the Ministry of Agriculture of Kametaro Toyama and his fellow agriculture experts from Japan.[20]

The Prince then proposed a number of practical measures for the advance of agriculture in Siam. The government should provide indirect encouragement to this sector, through the organization of agricultural fairs to promote improvements in crop strains and in agricultural implements, by the enactment of a patents law to provide protection (and thus encouragement) for agricultural practices and implements newly developed in the kingdom, and through the introduction of legislation to standardize weights and measures in Siam.[21] But, argued Prince Phenphat, the government would also have to act directly to advance agriculture, through the establishment of experimental farms which would not only develop more productive agricultural practices but which would also act as models and sources of information in more advanced techniques for the kingdom's cultivators, by the establishment of a draught-animal breeding station, and through the founding of an agriculture school. In order to prepare itself for such direct action, the Prince proposed that the government send students abroad to study agricultural science and also engage a European official to undertake an accurate statistical survey of the kingdom's cultivation and trade. Čhaophrayā Thēwēt fully supported his subordinate's views.[22] However, the King, although encouraging, saw in his son's letter simply 'preliminary ideas'.[23] These now needed to be thought through, to be worked into precise, practical proposals for governmental action.

In the years which followed, Prince Phenphat sought to act on his father's advice. Not all his initiatives found success. Most notably, although the Prince gave considerable time to the planning of experimental farms in Rātburī and Nakhǫn Pathom, neither developed into a permanent centre,[24] as presumably was envisaged. In fact, his major effort in these years appears to have been concentrated in the establishment of specialist agricultural education in the kingdom. January 1905 saw the founding in Bangkok of a Sericulture School, the principal objective of which was to train Siamese officials who could in time replace the Japanese engaged from the early years of the decade to initiate the administration's sericulture programme.[25] The following year the school was renamed the School of Agriculture, reflecting an expansion of its curriculum into other branches of cultivation. Then in November 1908, the School of Agriculture was amalgamated with the schools in the Survey Department and in the Canal Department to form the School of the Ministry of Agriculture. It is difficult to assess the extent to which this expansion in agricultural education in the later 1900s could have met the manpower requirements of the agricultural administration. The short-term result was that in the three years 1907–9, just nineteen students graduated from the government's agricultural school.[26] Of these, no less than fourteen were trained in sericulture and found employment in the administration's short-lived silk programme. The remaining five continued their training overseas, three being accepted at German colleges to study engineering (presumably these were officials from the Canal Department), while two went to the United States to study surveying. It is possible that in time the School of the Ministry of Agriculture would have shed its sericulture origins and produced many more graduates with a training directly relevant to the principal agricultural concerns of the kingdom. But in November 1909 Prince Phenphat, the main driving force behind this expansion in agricultural education, died of tuberculosis, at the age of 27.[27]

According to an informed contemporary observer, in the years immediately following the death of Prince Phenphat, the agricultural school 'went to pieces under bad administration'.[28] Towards the end of 1912, it was absorbed into the Civil Service College.[29] The latter had been founded by Prince Damrong in 1899 as the Royal Pages School.[30] Until 1910 it had been under the authority of the Prince's Ministry of the Interior, and indeed its students had been trained exclusively for that Ministry; but in that year, administration of the Royal Pages School had become the responsibility of the government as a whole, although until Prince Damrong's ministerial resignation in 1915 his Ministry continued to dominate the by now renamed Civil Service College. Thus, with the absorption of the School of the Ministry of Agriculture into the Civil Service College in 1912, the specialist role of the former was inevitably lost, an eventuality confirmed by the government's simultaneous dismissal of the European staff of the agricultural school.[31]

In reporting these events, the *Bangkok Times* added a distinctly sour assessment of the agricultural school's brief history.[32] '... the school has always led a somewhat precarious existence, and has never filled that place and position which agricultural schools occupy in other countries where agriculture plays an important part in the life of the people.' Specifically, the 46 students who had graduated from the school and been placed in the provincial administration had been unable to exert significant influence in advancing cultivation practices, for 'the potentialities of their help have not been realized by the Monthon [provincial] officials. [Indeed] ... these land officers are not at present asked to concern themselves with the advancement of scientific agriculture.' The *Bangkok Times* concluded: 'But what we are concerned to insist on is that experimental farms where scientific principles are worked out, are just as essential in Siam as in all neighbouring countries, and that this side of the Ministry's [of Agriculture] work ought as soon as possible to be emphasized.'[33]

It was argued earlier that in the establishment and direction of agricultural education in Siam from the mid-1900s, a primary concern of Prince Phenphat had been to ensure that the administration's agricultural officials would have at their command both a firm knowledge of the principles of scientific agriculture then being advanced in the Western world and a close familiarity with local cultivation practices and conditions. But this objective was not to be easily achieved, for reasons which were more fully revealed around the time of Prince Phenphat's death. At the end of September 1909, during the annual meeting of the provincial superintendent commissioners in Bangkok, the acting Minister of Agriculture, Čhaophrayā Wongsānupraphat, proposed the appointment of agricultural commissioners (*khāluang kasēt*) to, initially, three *monthon*.[34] Their responsibilities were to include carrying out inspection tours, assisting local cultivation and, perhaps most notably, submitting regular reports to the Ministry of Agriculture in the capital on agricultural conditions and developments in their area. The fact that it was proposed to appoint agricultural commissioners initially to only three *monthon* would suggest that the administration anticipated some difficulty in finding officials with the training and experience appropriate to this position; and indeed at the end of November 1909, Čhaophrayā Wongsānupraphat informed the King that he had made an appointment to just one *monthon*, Bangkok, and even that only with considerable difficulty.[35] By way of explanation, the Minister argued that were he to appoint as agricultural commissioner, a young official with knowledge only of European agricultural practices, the appointment would probably not be effective, for agriculture in Siam continued to be practised mainly with long-established, traditional methods which could be reformed only gradually and with subtlety. He had therefore turned for his first agricultural commissioner to an established official in the Ministry of Agriculture who had knowledge of modern cultivation methods but, perhaps more importantly, also a firm understanding of the traditional practices of the Siamese cultivator. This raised precisely the issue which had so concerned Prince Phenphat from earlier in the decade.

Čhaophrayā Wongsānupraphat's views prompted a more penetrating analysis from the King.[36] Chulalongkorn saw a government administration divided along generational lines. On the one side were

... the young officials who know nothing of the old practices but who are familiar with effective [modern] methods. However because they are unable to present their knowledge in an instructive manner—an inability derived from their [self-assured] natures—they are scarcely able to use that expertise to reform the old practices. They become irretrievably depressed. They can achieve nothing.[37]

On the other side of the divide were the older officials. Some were intent on reform, but as they did not have a knowledge of basic principles these officials were forced to proceed essentially by trial and error. Inevitably they groped their way forward, frequently to failure. In time they had become discouraged and increasingly infuriated. Given this divide, the primary task of the administration, argued Chulalongkorn, was to seek to combine indigenous and foreign practice, to fuse the old and the new. But to achieve this would necessitate breaking down a formidable barrier within the bureaucracy.

... there must be a way to bring together the younger and the more experienced officials—so that the more experienced will learn new practices from officials who are younger than they are without feeling, shamefully, that they are being taught by children. Similarly those young officials with extensive knowledge must be persuaded to learn from their seniors who have experience of local conditions and practices, yet without feeling that they are students of primitive old men who know nothing at all about cultivation. These two generations have failed to blend because of stubborn pride [on both sides].[38]

The measures taken by the Siamese administration in the early twentieth century to secure a substantial corps of agricultural specialists, and the attendant analyses of Prince Phenphat, Čhaophrayā Wongsānupraphat, and the King, have been considered at length simply to illustrate the complexity and intractability of the problems faced by the government in this important area. That corps could not be created simply by instilling in selected officials a knowledge of the principles of agricultural science and ensuring their familiarity with advanced cultivation practices then employed in the Western world. In the first place, at a time when state secular education was still very much in its infancy,[39] the number of young Siamese with the educational foundation that would enable them to undertake specialist instruction in agricultural science—indeed in any advanced technical field—was inevitably limited. It should be noted in this context that the agricultural school's curriculum included instruction in mathematics and in the sciences.[40] Second, as Prince Phenphat clearly recognized, in order to secure the advance of the kingdom's agriculture it was necessary to create a body of distinctive agricultural knowledge specifically appropriate to Siamese conditions; this could be achieved only through the painstaking adaptation of advanced Western agricultural methods by officials who had a close familiarity with local agrarian conditions and traditional cultivation practices. Yet, according to the King, in its attempts to fuse the Western knowledge acquired by the younger generation of officials and the extensive experience of indigenous conditions and practices held by their seniors, the administration was confronted by a major divide within the bureaucracy, a divide that had its roots deep in Siamese society and culture. And finally, it would appear that the effectiveness of the relatively few agriculture officials who were trained in this period was seriously diminished by the failure of the provincial administration to appreciate the importance of their expertise.[41]

Faced with these complex and intractable problems, the Siamese administration was inevitably constrained both in the establishment of programmes of agricultural experiments and trials and in the subsequent diffusion of more advanced cultivation practices through the principal agricultural districts of the kingdom. With respect to the latter, it is evident that during this period the government was simply unable to place a sufficient number of agriculture officers in the rural districts to effect an advance in cultivation practices—whether or not their expertise was fully exploited. It was noted earlier that at the end of the 1900s the administration, with considerable difficulty, made just one appointment to the pivotal post of agricultural commissioner. In the early years of the following decade the provincial establishment of the Ministry of Agriculture appears to have expanded quite substantially,[42] presumably in large part as a result of the emergence of the initial graduates from the agricultural school. Yet it must be added that a recommendation of the provincial superintendent commissioners made as early as 1910 that agriculture officers be posted in all districts in the kingdom could not be fully implemented until after the Pacific War.

With respect to the establishment of agricultural research within Siam, it was noted earlier that the creation of an experimental farm had been seriously considered by the administration towards the end of the 1890s; that an attempt by Čhaophrayā Thēwēt in the opening years of the 1900s to establish Nāi Čharoen as the foundation for this project had broken down as Nāi Čharoen himself had failed; but that Kametaro Toyama, Prince Phenphat, and Čhaophrayā Wongsānupraphat had continued to argue for this initiative throughout the remainder of the decade.[43] But in fact, whilst the administration failed to bring itself to action, some members of the Siamese élite began to undertake agricultural experiments on their own initiative and in a private capacity. Prominent here were members of the Sanitwongse family. From the 1890s Sai Sanitwongse, a founder partner of the Siam Land, Canals, and Irrigation Company and a major Rangsit landowner, along with his eldest son, Suvabhan, began experimenting with various forms of agricultural machinery, including steam ploughs and powered rice-threshers, on their Rangsit holdings.[44] Suvabhan also had an important interest in improving the quality of the rice seed planted by the kingdom's cultivators, and his work here included the organization of an agricultural exhibition held in Thanyaburī district in 1907 at which local farmers were invited to submit samples of their crop in a rice seed contest. This was an initiative which the government soon took up for its own. With respect specifically to the establishment of agricultural research, the government was finally moved to action in the mid-1910s. The principal stimulus here appears to have been the administration's concern from the early years of that decade that the kingdom's rice was now facing increasingly strenuous competition in international markets from Burmese and Cochin-Chinese rice.[45] As was noted in the previous chapter, it was principally this concern that also prompted Prince Rātburī in 1913 to revive proposals for large-scale irrigation works.[46] Of particular concern in the present context was the apparent practice in the Siamese rice trade of mixing different grades of grain within individual consignments.[47] One immediate consequence of this practice was that each consignment of Siamese rice came to be sold at a price set by the lowest grades within it. Moreover, according to one source, on taking delivery of the kingdom's rice, traders would mechanically sift consignments into their various grades, sell the highest grades as premier Indian rice, and dispose of the poor quality balance as Siamese rice[48]—so further damaging the reputation of the kingdom as an élite rice producer. According to this same source, this experience brought home to the administration the importance of careful seed selection and of securing improvements in seed quality if rice cultivation and export were to progress; this, in turn, firmly established the demand for an experimental rice farm.

The farm was eventually founded in 1916/17.[49] Located in the Rangsit district, it concentrated on pure line selection of rice seed varieties, experiments with agricultural machinery and, from the early 1920s, fertilizer trials and soil testing. Feeny has argued that the siting of the experimental rice farm specifically in Rangsit once again reveals the powerful influence of the Rangsit landowners within the Bangkok élite.

> The land in this area was largely owned by the Bangkok élite, who were concerned with the decline in paddy yields and were able to arrange for the location of the social research facilities in the area where they would be able to benefit the most. While the Rangsit area was a major rice-producing area, Ayuthia was a larger producer and might have been a more logical location for a rice experimental farm.[50]

Feeny further indicated that the administration was willing to invest in agricultural research in this period only to the extent that the private interests of the élite would be the principal beneficiary.[51] But, as with Feeny's analysis of the Siamese administration's irrigation policy (considered in the previous chapter), whatever the logical attractions of his argument, no primary documentary evidence is offered in support of the view that private élite interests were the crucial influence in this area. Indeed, it is possible to suggest sound administrative and agricultural reasons why the Rangsit district might have been chosen as the site for the experimental rice farm: it was only a short distance from the Ministry of Agriculture in Bangkok; it was a district of notably intensive cultivation and consequently an area where the problem of declining yields was particularly acute; and it had the most developed water-control facilities in the kingdom.[52] As a final comment on Feeny's arguments here, it might be added that in view of the administration's oft-expressed concern from the beginning of the 1910s that the kingdom's rice now faced increasingly strenuous competition in international markets, it seems unreasonable to suggest that the government was prompted to establish the experimental rice farm only by a wish to protect the private landowning interests of the Bangkok élite.

The Siamese administration pursued two further agricultural initiatives in these years. As was noted earlier, in 1907, Suvabhan Sanitwongse had organized an agricultural fair in Thanyaburī district which prominently included a rice seed contest. The apparent success of this event encouraged the government itself to organize a fair in Thanyaburī the following year, and again on a much larger scale in Bangkok in 1909.[53] This in turn led the government to organize the First Annual Exhibition of Agriculture and Commerce, again held in Bangkok and opened by the King in April 1910. The exhibition, which included a wide variety of crop competitions and demonstrations of mechanized agricultural equipment, attracted tens of thousands of visitors. A second exhibition organized along the same lines was held in 1911, but it attracted considerably less visitors and participants, partly because it coincided with the period of national mourning following the death of King Chulalongkorn. No further national (or indeed local) agricultural exhibitions were held in this period. The second initiative was the publication by the Ministry of Agriculture of a monthly journal, *Prakōp kasikam* (*Agriculture*), the first issue of which appeared in 1911.[54] It was intended primarily to provide provincial officials with a wide range of information relating to agriculture and trade. However, although each issue contained valuable reports on crop prices and on agricultural conditions in the kingdom, its articles appear to have involved general exhortations in pursuit of economic change (the journal was particularly concerned to advance agricultural diversification within the kingdom's rural communities and to encourage the purchase of locally-produced goods in preference to imported manufactures), rather than have provided detailed practical information and advice on more productive cultivation methods. *Prakǫp kasikam* ceased publication in 1913.

It is difficult to assess the impact of the agricultural fairs and exhibitions, of *Prakǫp kasikam*, and of the Rangsit experimental rice farm on the rice economy of Central Siam in this period. On the one hand, there are some contemporary references to notable advances in cultivation practices and in rice seed quality. For example, at the opening of the second exhibition in 1911, the Minister of Agriculture, Čhaophrayā Wongsānupraphat, remarked that the rice seed samples submitted for competition in that year were of a consistently higher quality than those submitted for the exhibition of 1910, and then added, 'our farmers have begun to understand the practices of rice seed selection and to understand improved methods of rice cultivation'.[55] In addition, one source has suggested that by the mid-1920s the rice varieties developed at the experimental rice farm in Rangsit were highly regarded by the international rice market.[56] This view is reinforced by the fact that Siamese rice varieties were notably successful at the World Grain Exhibition and Conference held in Regina in 1933.[57] But against this evidence, and outweighing it, is the bald fact that through the period from 1910 to 1940 paddy yields per hectare for the kingdom as a whole had a distinct tendency to fall.[58] It is possible that the Siamese administration was simply too tardy in establishing an experimental rice farm and too hasty in abandoning the agricultural fairs and its agricultural journal; perhaps the fairs and the journal were not very effective mechanisms for diffusing more advanced rice cultivation practices through the agricultural districts of the Central Plain and that this process really demanded the establishment of a substantial corps of agriculture officials in the field; it is possible that, in any circumstances, far more time was required for agricultural advances such as those apparently achieved at the Rangsit experimental farm to be made effective through the rural communities of the kingdom.

An important perspective on the failure of the Siamese administration of the early twentieth century to secure any significant advance in rice cultivation practices or inputs in the Central Plain is gained by a brief examination of the programmes of agricultural reform pursued in other rice-important economies in Asia in this period. Thus, in Cochin-China, the French administration's practical commitment at the very least to improving rice seed inputs appears to have been distinctly weak. It was not until 1927 that a central rice station, to distribute improved seed varieties through the rice districts, was established at Can-tho in the trans-Bassac.[59] In 1930, L'Office indochinois du riz was created to undertake research and practical trials in rice cultivation, and to propagate improved inputs and practices.[60] The British administration in Burma brought an earlier and stronger commitment to this basic work. Following the establishment of a separate Agricultural Department in 1906, a central experimental farm was created at Mandalay to secure advances in cultivation in Upper Burma.[61] In 1914 another experimental farm was established at Hmawbi, a short distance north of Rangoon, concerned almost exclusively with rice culture. Further experimental farms were created from the mid-1920s, each concentrating on rice. The work of these farms followed familiar lines—the selection, multiplication, and distribution of higher quality rice seed; the study of soil conditions and fertilizer inputs; and the development of improved agricultural implements and cultivation methods. The American administration in the Philippines appears to have been comparably concerned to improve agricultural inputs and practices. A Bureau of Agriculture was established as early as 1901, to undertake experiments that would improve yields, develop new crops, and eradicate plant and animal diseases and to propagate modern agricultural practices among Filipino farmers.[62] In addition, the opening years of American administration saw the establishment of a national agricultural college at Los Baños, supplemented by a number of high school-level agricultural schools throughout the country.[63] But, crucially, even in those states in South-East Asia where the administration made a markedly greater investment in agricultural advance in the early twentieth century than was achieved in Siam, the impact on rice cultivation practices and inputs was, as in Siam, negligible. Thus, for example, even as late as 1929-30, less than 2 per cent of the total area under rice in Burma was planted with improved varieties; or again, there was a gradual decline in the production of rice per acre in most regions of Burma during the first decades of the twentieth century.[64] Indeed it would appear that in the period 1928-32, the annual average rice yield per acre in Burma was significantly below that in Siam, although considerably above that in Indo-China.[65] In brief, government commitment and investment alone appears not to have been sufficient to raise rice yields in South-East Asia in this period.

It is at this point that the Japanese experience may be instructive. Between 1880 and 1930 land productivity (per hectare of crop area) in Japanese agriculture rose by 78 per cent; labour productivity (per workday) rose by 87 per cent.[66] The explanation for these major increases remains an issue of considerable controversy among economic historians of Japan, but a number of points may be sketched in.[67] It might first be noted that in the initial phase of agricultural advance (the period to the early twentieth century), the volume of government resources allocated to agricultural research and development was very modest, sufficient only for simple field experiments comparing seed varieties or cultivation practices to be undertaken. In fact, the initiative in agricultural innovation in Japan in this period was taken primarily not by government research staff and technicians but by the cultivators and landlords themselves. Crucial here were those landlords who personally farmed part of their holdings (the *gōnō*), and among this class, the *rōnō* (veteran farmers). From the later 1870s experienced cultivators began to develop improved rice production techniques and inputs, including 'salt-water seed-selection, short-strip seed-bedding, straight-line replanting, water management, [and] compost-making and its application'.[68] Innovations in these areas tended to be specific to the locality in which they were developed; but their modification under trial at the government agricultural experiment stations frequently gave them an applicability throughout Japan. The more advanced agricultural practices and inputs were diffused primarily through agricultural societies organized principally by *gōnō* farmers and landlords, strongly supported by the central and prefectural governments. Stated broadly, it has been argued that

Rapid agricultural growth in the latter half of the Meiji era [1868–1912] was based on a backlog of indigenous technological potential, previously dammed by feudal constraints, which was released by the reforms of the [Meiji] Restoration.[69] The interaction among farmers, agricultural scientists, and agricultural-supply firms was effective in exploiting the potential. The initiative of innovative farmers and landlords, together with the proper guidance of the government, was the key to the effective interaction.[70]

The argument that the initial phase of agricultural advance in Meiji Japan was achieved principally through the initiative of innovative farmers, although firmly underpinned by the central and prefectural governments, leaves a crucial question. Why were Japanese rural communities so innovative in this period? By implication, why were Siamese cultivators not comparably innovative? This is a deeply complex issue. It is possible that in the early twentieth century, rural Siam simply did not possess the backlog of indigenous technological potential that was so crucial in near-contemporary Japan. It is certainly the case that in this period Central Siam did not possess the comprehensive irrigation and drainage system essential for the full exploitation of more advanced agricultural inputs and practices, whilst Meiji Japan was notably well endowed in this respect.[71] It is also possible that the shortage of cultivable land relative to population in Japan demanded agricultural innovation in a way which did not occur in Central Siam, where, in the early twentieth century, there was abundant land for the cultivating population. Beyond this, the explanation must reside largely in the major differences in rural economic and social structure as between Japan and Siam. Important here are the observations that by the Meiji period 'the small-scale [increasingly rented] family farm of around one hectare of cultivated land [was] the dominant mode of agricultural production' in Japan[72] and, perhaps more notably, that the rural communities of Japan in this period had a powerful organizational capacity:

> Highly stringent feudal rules on various facets of rural life, including the collection of feudal tax in kind, had been enforced through the village leaders, which had the effect of training local leadership. The organizational capacity of farmers had also been promoted by the need to cooperate in constructing and maintaining local infrastructures, above all, the irrigation system.[73]

This analysis need not be pushed further here, for the essential argument has been established. If the foundation of increased agricultural productivity in Siam in the early twentieth century, as in Meiji Japan, lay primarily in the structure of social authority and economic power within the kingdom's rural communities, then inevitably this was an area which was in large measure beyond the intervention of the Siamese administration of that period.

II

The concluding section of this chapter will consider one further aspect of the Siamese administration's perception of and policies toward agricultural change in the early twentieth century—the response of the administration towards the rice economy in distress, specifically its reaction to what was widely seen as a sharp recession in the kingdom's rice trade in the second half of the 1900s.

In early 1906, after receiving reports that cultivators were leaving Rangsit in substantial numbers, Prince Damrong sent three of his officials to the area to investigate.[74] In their subsequent report, the officials identified two main causes of the out-migration.[75] First, they noted that over the years substantial quantities of silt and other debris had been allowed to build up in the Rangsit canals to the point where those deposits now hindered water-borne transportation through the district and, perhaps more seriously, prevented the adequate irrigation and drainage of the adjacent rice-fields. Responsibility for the maintenance and repair of the canals was said to lie with the Siam Land, Canals, and Irrigation Company but, continued Prince Damrong's officials, the company had shown little interest in this work in the past. In their consideration of the second principal cause of the out-migration from Rangsit, the officials were strikingly blunt:

Most of the paddy land [in the Rangsit district] is rented from absentee landlords who live in Bangkok. [Whilst the district was still being opened up, the landlords treated the cultivators equitably.] But as soon as the former felt confident that the farmers wished to settle permanently they demanded major increases in rents. [Of course] by this stage any farmer would have found it very difficult to remove his buildings and belongings ... and consequently in certain areas extraordinarily high rents were enforced. [Moreover] once farmers have begun ploughing and planting they are summoned to the landlord and forced to enter into contract with him. [In these circumstances] the farmers have no choice but to agree to the owners' demands.[76]

The Ministry of the Interior officials continued:

In renting paddy land farmers are required to pay the full charge according to the size of the plot leased to them, for there exists no modified form of rent. In the event of the rice crop being damaged either by flood or drought farmers are still obliged to pay the rent in full since landowners will not reduce it.... In addition since most of the land is owned by absentee landlords, farmers are further exploited by the landowners' local agents.... [The latter] constantly demand the full payment of rents, showing no sympathy for the farmers' plight.[77]

This report was sent for comment to Suvabhan Sanitwongse, manager of the Siam Land, Canals, and Irrigation Company and a prominent Rangsit landowner.[78] Suvabhan confidently defended his interests. He countered the allegation that his company had failed to maintain the Rangsit canals with the argument that the fault in this respect lay principally with the cultivators themselves who freely tipped their refuse into the canals and who carelessly allowed their water-buffalo to trample down the canal banks. The authorities had done little to curb this damaging behaviour. On the question of rents, Suvabhan offered the explanation that the Rangsit landowners had been so shaken by a major increase in land tax rates introduced in 1905 that they had had insufficient time to consider land rent reductions; but once their tenants had become agitated, reductions had followed. Suvabhan concluded by outlining his own explanation for the out-migration from Rangsit. The cultivators there had recently suffered two very poor harvests which had left them seriously in debt; they had then had to face demands for land rent, irrigation dues, boat fees, and land tax. Inevitably, therefore, they had left the district in large numbers.

The report by the three officials of the Ministry of the Interior provided an admirable basis for a wide discussion through the administration of the problems then confronting the cultivators in the important Rangsit district; and in making his response, Suvabhan Sanitwongse, although understandably bounded by personal interests, had taken up the officials' challenge in a serious manner. Yet when these papers came before the Council of Ministers towards the end of May 1906, it was agreed that clearing and cleaning the Rangsit canals would in itself be sufficient to restore a measure of prosperity to the district.[79] At this time the Siamese administration appears to have been unwilling to confront the awkward issues which had been raised by the officials of the Ministry of the Interior and by Suvabhan—the nature of the landlord–tenant relationship in Rangsit; the burden of taxation imposed on the rice cultivator; and the essential inadequacies of irrigation and drainage in the Rangsit district.

However, some three years later these issues became a matter for keen debate among the English-reading public of the kingdom. In mid-September 1909 the *Bangkok Times* reported that a large number of cultivators, mostly from the Rangsit district, had come into Bangkok to petition the Minister of Local Government for remission of the land tax. The editor commented that 'the chief reason for the inability of the small people to pay their taxes is to be found in the very high rents which they have to pay'.[80] This comment provoked a large correspondence. A number of writers directly challenged the editor's statement. One correspondent, 'Paddy Planter', claimed that during the current recession 'many of the landlords have taken very little, or no, rent, while we are obliged to pay nearly the full amount of land tax'.[81] A later writer, 'A Landlord', reinforced the last point, arguing that the government's revenue officials had shown no compassion in their collection of the land tax.[82] But the *Bangkok Times* correspondence rapidly went beyond the issue of the relative burdens of land taxation and rent payment. The depressed state of cultivation in Rangsit was said to be the result of: the high exchange value of the baht which had made it difficult for Siam's rice to compete with rice shipped out of Rangoon and Saigon, and had indeed 'practically killed export to Hongkong';[83] the recent enforcement of military conscription which had temporarily broken up many cultivating families;[84] the recent series of heavy monsoons which had caused considerable flood damage to the rice crop.[85] One writer, signing himself 'Critic', argued that in Rangsit, 'the real evil ... is the silting up of the canals. The Government ... neither ends the Company's control nor compels the canals being put in good condition. Instead it closes down the Irrigation Department for most purposes. To my mind the blame lies at the door of Government for the steady deterioration of the Rangsit district.'[86]

These diagnoses led naturally to proposals for alleviative action. Thus there were calls for the devaluation of the baht, for the introduction of selective exemption from conscription, and for the construction of effective irrigation and drainage works in the principal rice districts. However, note should also be made of three further alleviative measures proposed in the *Bangkok Times* correspondence columns. Suvabhan Sanitwongse argued for an important change in the ploughing cycle in the Central Plain, a change which would require the rice cultivator to use mechanical ploughs in place of his existing wooden implements.[87] One correspondent sought the suppression of gambling,[88] long felt to be a major threat to the prosperity of the Siamese cultivator. Third, and most important, there were calls for the establishment of a government land bank which, with branches in the provinces, would advance loans to cultivators at a moderate rate of interest.[89] This proposal was much favoured by the *Bangkok Times* itself.[90]

The discussion in the correspondence and editorial columns of the *Bangkok Times* as to the probable causes of and possible solutions for the Rangsit distress took place at a time when the Siamese administration itself, and most notably the Minister of Agriculture, had just embarked on its own lengthy consideration of those same issues. The agricultural administration's deliberations, which extended from August 1909 to December 1910, are of considerable interest, in part because of the range of ideas they produced but more particularly because they offer a valuable insight into the attitudes of the Siamese administrative élite towards the rice cultivator and his problems during this period.

In mid-1909 a group of distressed farmers in Thanyaburī district met to discuss the crisis with which they were faced.[91] Heavy rains in 1908 and 1909 had severely damaged the rice crop, leaving them with nothing to eat, no seed for planting, and no cash to meet their land tax and rent demands. Their landlords had taken action to seize their water buffalo in compensation. The farmers decided at their meeting to petition the government for loans to enable them to continue cultivation. The acting Minister of Agriculture, Čhaophrayā Wongsānupraphat, discussed this request with the government's General Adviser, Westengard. The latter argued that the administration should be prepared to make loans to the Thanyaburī farmers; and he also took the opportunity to send to the Minister a copy of a memorandum on the Agricultural Bank of Egypt, prepared earlier in the year by W. J. Archer at the Siamese Legation in London.

But Čhaophrayā Wongsānupraphat was unsympathetic.[92] He was opposed to the establishment in Siam of an agricultural bank along the lines of the Agricultural Bank of Egypt on the grounds that there were major differences in agricultural conditions between the two countries. He pointed out, for example, that while the Egyptian cultivator was burdened with debt and was required to work poor soils, the same was not true of his counterpart in Siam.[93] More interestingly, the Minister noted that whilst Egypt had of necessity extensive water-control works which ensured a fine cotton crop (and thus presumably the ability of the Egyptian farmer to meet his debt obligations) in almost all years, in Siam the rice cultivator (and thus his income) was at the mercy of the monsoon.[94]

It might be argued that this comparison, more than having a bearing on the issue of an agricultural bank, simply re-emphasized the urgent need for adequate irrigation and drainage works in the Central Plain of Siam. But if so, there is no indication that Čhaophrayā Wongsānupraphat wished to draw that conclusion. The Minister's failure to acknowledge the importance of adequate irrigation and drainage is made all the more surprising by his willingness to accept the argument of the Thanyaburī farmers that they had been brought to their present distressed state by the very heavy rains of 1908 and 1909 which had destroyed virtually all their rice crop; for again he did not draw the obvious conclusion that that destruction would have been substantially reduced if the Rangsit district had had effective drainage. Rather, Čhaophrayā Wongsānupraphat was determined to argue that the rice farmers should have prepared themselves to withstand a failure of their crop.

This argument was to be constantly restated by the Minister of Agriculture over the following months. In essence, Čhaophrayā Wongsānupraphat saw the Siamese cultivator as being irresponsible, wasteful, and lacking perseverance. For example, he suggested that it was in the farmer's character to dissipate, frequently by gambling, virtually all the surplus earned in good crop years, so leaving no reserves to sustain himself should a later crop fail. The cultivators were now impoverished not because they had insufficient capital to undertake farming. They were impoverished, declared the Minister, primarily because they had not learnt the crucial importance of being frugal. And it was in order to instruct the Siamese cultivator in that virtue that Čhaophrayā Wongsānupraphat proposed the creation of a government savings bank, with its branches sited in provincial and district treasuries. It was clearly not intended that the bank would channel additional funds into the agricultural districts. The Minister held that the capital requirements for rice cultivation in the Central Plain were modest; and he opposed the government providing loans for the cultivator. Rather, the proposed bank would simply receive deposits from the rural population, and in this way inculcate in them an awareness of the importance of thriftiness and of making provision for bad times.

Čhaophrayā Wongsānupraphat further proposed a number of measures that would assist rice cultivation directly.[95] He argued that the terms and conditions upon which Lao labourers from the north-east of the kingdom were hired for seasonal work on the rice holdings of the Central Plain should be firmly regulated. This would ensure for the rice farmer a secure body of additional labour during the peak demands of the agricultural cycle, and an agreed wage for the labourer from the north-east. The Minister called for improvement in the quality and care of the kingdom's draught animals and for an increased use of machine-driven ploughs and rice threshers. He argued that the government should waive the collection of land tax on newly-cleared holdings, and should allow at least partial remission of land tax in those cases where the rice crop was seriously damaged by abnormal rains. Čhaophrayā Wongsānupraphat proposed the appointment of agricultural commissioners at the *monthon* level and of cultivation officers right down to the village level. He called for provincial administrations to provide increased protection for the possessions and lives of cultivators, in particular protection from the depredations of cattle thieves. He also proposed the creation of a number of experimental farms.

Čhaophrayā Wongsānupraphat's proposals were presented in full to the Council of Ministers on 20 August 1909.[96] They provoked very little discussion, probably because the sheer number of ideas thrown out by the Minister of Agriculture discouraged consideration of his programme as a whole, whilst paradoxically the absence of precise proposals for government action left the assembled ministers with little to discuss. Prince Damrong's principal contribution—to contrast the well-maintained rice-fields in the Sāen Sāep district where the majority of the farmers were owner–occupiers with the seriously deteriorating conditions in Rangsit where there was a high degree of tenancy—was a provocative observation, carrying echoes of the earlier report by the officials from the Ministry of the Interior. But no one responded. In closing the discussion, the King approved Čhaophrayā Wongsānupraphat's 'programme', adding that as its implementation would require the assistance of the Ministries of the Interior, Finance, and Local Government, he should now consult with them.[97]

In fact, Čhaophrayā Wongsānupraphat's proposals next came under discussion not in inter-ministry consultations as the King had wished but at the annual meeting of the provincial superintendent commissioners held in Bangkok at the end of September 1909.[98] At a lengthy session devoted to a consideration of agricultural conditions, the superintendent commissioners heard the acting Minister of Agriculture outline the proposals which he had placed before the Council of Ministers the previous month. They were prepared at this point to approve only his proposal for the appointment of agricultural commissioners. Implementation of his other measures—the creation of a government savings bank, the regulation of the terms and conditions upon which Lao labourers were hired, the improvement in the quality and care of the kingdom's work animals and the increased use of powered machinery in rice cultivation, the granting of land tax exemptions and remissions—should wait until those officials were installed in the provincial administration. As was noted earlier, the first agricultural commissioner was appointed towards the end of that year.

Although the King approved the decisions taken at the September meeting of the provincial superintendent commissioners,[99] he was undoubtedly concerned that Čhaophrayā Wongsānupraphat had still to discuss his numerous schemes in detail with the Ministers of the Interior, Finance, and Local Government and, as a consequence, was still some considerable way from reworking his preliminary ideas into precise, practical proposals for government action. It was presumably in order to encourage his Minister in this direction that around this time the King suggested that he seek the views of Sai Sanitwongse, a founder partner of the Siam Land, Canals, and Irrigation Company and a major Rangsit landowner (and indeed the father of Čhaophrayā Wongsānupraphat), on the current problems of rice cultivation and water control in the Central Plain, and that he convene a meeting of 'wise and learned men' to hear their ideas as to the measures necessary to revive the depressed rice economy.[100] Rather surprisingly, Sai could provide no insight into these problems, despite the fact that he had recently spent some considerable time in the rice districts. Therefore, still seeking guidance, Čhaophrayā Wongsānupraphat then called a meeting of Rangsit landowners at the Ministry of Agriculture in early December 1909. It is not clear to what extent this gathering was intended to satisfy Chulalongkorn's wish for a meeting of 'wise and learned men'.

Opening the December meeting, Phrayā Srīsunthǫnwōhān, a senior official in the Ministry of Agriculture, explained to the assembled Rangsit landowners that the administration was seeking their views on the measures required to restore rice cultivation, specifically in their district.[101] The landowners responded with a barrage of opinions, many familiar from the recent correspondence in the *Bangkok Times* and from Čhaophrayā Wongsānupraphat's analyses over the preceding months. The baht was overvalued; the administration had been insufficiently flexible in granting land tax remissions; the labourers from the north-east were unreliable and dishonest; there was an urgent need to inoculate work animals against rinderpest and anthrax. But two issues in particular concerned the landowners. The first was that the Rangsit canals urgently required clearing and cleaning. Second, a number of landowners referred to the ruinous influence of gambling on the rice farmer, noting specifically that if landlords advanced capital to their tenants to assist cultivation, not infrequently those funds were largely lost in playing the lottery. Consequently, there were calls for the immediate removal from Rangsit of the facilities for participating in the lottery. More significantly, one speaker argued that, in the circumstances, it would be inadvisable for the government to establish an agricultural bank advancing capital to the cultivator, for those funds too would soon be lost in gambling.

In reviewing these arguments with the King, Čhaophraya Wongsānupraphat once again demonstrated his facility in throwing out general ideas and his inability to work them into precise proposals for government action.[102] Responding to the demands that the Rangsit canals be cleaned, the Minister now proposed the creation of agricultural associations in that district, formed from among the cultivators there. The principal responsibility of the associations, at least initially, would be to protect the canals in their designated areas apparently from the damage caused by cooling off draught-animals and from the unrestricted tipping of debris. The King replied curtly, instructing Čhaophraya Wongsānupraphat to discuss his preliminary ideas with the other appropriate ministries to produce more substantial proposals.[103] Yet, once again the full import of that simple instruction appears to have escaped the Minister. Towards the end of January 1910, Čhaophraya Wongsānupraphat received a petition from 57 Rangsit cultivators.[104] Their complaints—falling rice prices; the burden of an increase in land tax rates, imposed from 1905; the movement of rice barges being severely hindered by the unchecked silting of the Rangsit canals; rice crops seriously damaged by flood; the unreliability (indeed severe shortage) of Lao labourers—were all too familiar. In forwarding this petition to Chulalongkorn, Čhaophraya Wongsānupraphat added that he had outlined the measures necessary to alleviate the cultivators' distress in his earlier letters; and he reported that, following the King's instructions, he had sent copies of those papers to the other ministries which had a responsibility in this area.[105] Prince Damrong had indicated his approval; the other ministers had yet to respond.

At this the King's patience broke. In reply he led Čhaophrayā Wongsānupraphat step by step through the procedures by which it was intended the administration would formulate precise measures for implementation.[106] Those procedures required, crucially, that the minister with the principal responsibility in a particular area must think and plan in strictly practical terms (rather than simply throw out general ideas), and that all proposed measures be rigorously discussed with (rather than vacuously approved by) the other appropriate ministries. However, even this lecture on elementary administrative procedures appears to have left little impression on the Minister of Agriculture. Some six months later, towards the end of September 1910, the provincial superintendent commissioners met for their annual conference in Bangkok, and devoted two sessions to a consideration of agricultural problems.[107] Particular attention was paid to the possibilities for agricultural diversification in Siam and to Čhaophrayā Wongsānupraphat's proposal for the formation of agricultural associations in the rural districts, but once again the discussion appears to have been conducted in very general terms. In fact, the superintendent commissioners concluded that further consideration of these and comparable proposals to assist cultivation should wait until after the Ministry of Agriculture had received reports, then being prepared, on the agricultural conditions, trade, and economic potential of each *monthon*. That further consideration was to be undertaken primarily by a committee composed of the Ministers of the Interior, Public Works, Finance, Local Government, and Agriculture. The Minister of Agriculture was to be responsible for preparing proposals for consideration by the committee as a whole. The committee never met.[108]

A little over two months after this meeting of the provincial superintendent commissioners, Čhaophrayā Wongsānupraphat submitted a further memorandum which, in its analysis of the causes of the agricultural recession and its proposals for ameliorative action, was perhaps even less carefully considered than his earlier contributions.[109] In his review of the causes of the recession, the Minister trod familiar ground—the recent increase in land tax rates, the breakup of cultivating families by compulsory conscription, the high exchange value of the baht, and the recent series of poor harvests had each played its part. But critically, in Čhaophrayā

Wongsānupraphat's view, those adverse circumstances had been imposed on a people 'of thriftless and uneconomical habits, who do not take the least care to save for bad times', and who were 'naturally found ... absolutely unprepared with deplorable consequences'.[110] These weaknesses in the character of the Siamese cultivator, the Minister argued, '[had] been mostly brought about by long years of the gambling habit which is ingrained in their bones'.[111]

Čhaophrayā Wongsānupraphat's starkly unsympathetic and patronizing attitude towards the kingdom's cultivator was evident throughout. Thus, after listing some measures that would assist the rice farmer—the improvement of agricultural implements, in methods of draught-animal care, and in cultivation practices—he added: 'all these should be practically demonstrated to them until their minds thoroughly grasp that these are the best methods'.[112] And at a later point he observed that once the farmer (with government assistance) began to secure a surplus from cultivation, he 'will doubtless bury [it] in the ground'.[113] With such perceptions it was perhaps inevitable that Čhaophrayā Wongsānupraphat came to assume that a primary objective of the administration's agricultural measures would be to 'improve' the character of the cultivator. Thus, in responding to a report that European capitalists in alliance with influential Siamese were intent on securing sole control over the kingdom's rice-mills (and to the reality that Siam's rice trade was already in the hands of foreign merchants), the Minister outlined the following strategy:

By the direction and control of the Government, each community or district has to combine, and their united capital invested in rice mills in the rice market in Bangkok, and by that means, mill their own rice. It will be a lesson to them to save their surplus from their earnings, and to reinvest it in their own business, for increasing their own wealth and safety, and the most important of all, to learn to unite and cooperate for their own individual and the national interests under the fostering care of our paternal government.[114]

There is perhaps little need to add that this proposal, as presented, left untouched a number of formidable practical questions; for here, as so often in the past, Čhaophrayā Wongsānupraphat simply swept on to still more ambitious schemes. The passage quoted above continued:

By appointing suitable agents in all the [foreign] markets where our produce is sold, and by the chartering of steamships, which can now be had at a cheap rate, on account of severe competition all over the world, a considerable increase can be made in the profits. When prosperity comes ... we can buy these vessels for the rice trade of the markets close by. But this, by the proper support of the Government, could be developed into a national line to Europe and other countries where our produce may be in demand, and by its means train our naval reserve.[115]

The Minister was now in full stride. The government would have to consider the establishment of a national savings bank and an agricultural loan bank; an attempt should be made, partly by remitting local taxation, to encourage a measure of import-substituting production. But not once did Čhaophrayā Wongsānupraphat pause to consider, in practical terms, how any one of these proposals was to be put into effect.

Perhaps the most outstanding feature of the Siamese administration's consideration of the agricultural recession of the second half of the 1900s—from the report prepared by the Ministry of the Interior officials in 1906 to the *Bangkok Times* correspondence, the meetings of the provincial superintendent commissioners, the memoranda of Čhaophrayā Wongsānupraphat, and the meeting of Rangsit landowners in 1909-10—is that it produced no significant practical response from the government. Undoubtedly, part of the explanation here lies in the fact that, despite the length of the discussion, no clear consensus emerged as to the principal causes of the distress. Without plunging back into that argument, it may be suggested that a fundamental cause of the agricultural recession specifically in Rangsit was the sequence of heavy monsoons in the later 1900s which overwhelmed the inadequate and deteriorating drainage facilities in that part of the lower delta.[116] Unfortunately, Čhaophrayā Wongsānupraphat's rigid conviction that the gambling habit had drained the Siamese cultivator of financial and moral strength appears to have prevented him from accepting the full importance of adequate irrigation and drainage in maintaining the prosperity of the rice economy. In any event serious consideration of this issue was made virtually impossible by the government's decision in early 1909 to abandon all proposals for large-scale irrigation works. Furthermore, Čhaophrayā Wongsānupraphat's eagerness to condemn the Siamese cultivator as irresponsible, wasteful and irresolute, and his essential lack of sympathy for their distress (seen in his argument that the Thanyaburī farmers should have prepared themselves, by saving from good years, to withstand failure of their crop), virtually ruled out an impartial, precise assessment of the subsidiary causes of the recession. Thus, the Minister apparently made no serious attempt to determine the actual impact of the increase in land tax rates on the financial position of the cultivator; how seriously cultivation had been disrupted by compulsory conscription; or the precise extent to which the high exchange value of the baht had restricted rice exports. Nor did he (or indeed the administration as a whole) give serious thought to the observation contained in the Ministry of the Interior report in mid-1906 that the Rangsit cultivators were the victims of harsh and exacting landlords. The crucial argument here is that if the Siamese administration could not evolve an agreed, accurate explanation for the distress in the agricultural districts, then clearly there was no basis for effective ameliorative action.

There remains the formidable problem of accounting for the inability (or unwillingness) of Čhaophrayā Wongsānupraphat to work his numerous, often imaginative, initial ideas into practical proposals for administrative action. The explanation for this must lie essentially in the Minister's own character, for he has been described by another writer as ambitious, tactless, and ignorant.[117] However, the more intriguing question must concern the failure of the other ministries to react to Čhaophrayā Wongsānupraphat's inadequate procedures, particularly in view of the sharp criticism from the King himself. The ministries' inaction may simply represent one aspect of the near absence of unity and co-operation within the Siamese administration in the early twentieth century, frequently noted by contemporary observers.[118] Alternatively, it is possible that, despite the numerous meetings and lengthy memoranda, those ministries were not greatly disturbed by the failure of Čhaophrayā Wongsānupraphat to initiate practical measures to relieve the distress in Rangsit. For if the Siamese administration had been profoundly troubled by that distress, then it could be anticipated that either the King would have acted still more forcefully against Čhaophrayā Wongsānupraphat's failings or the other ministers with responsibilities in this area would have moved to take up and develop the ideas which flowed so freely from the Minister of Agriculture.

1. Chulalongkorn to Čhaophrayā Thēwēt, 30 June 1900, NA r5 KS 10/1.
2. Phrayā Suriyā to Čhaophrayā Phātsakǫrawong, 13 May 1900, NA S.Th 21.2/12.
3. Phrayā Wisut to Chulalongkorn, 29 June 1900, NA r5 KS 10/1.
4. Čhaophrayā Phātsakǫrawong to Prince Sommot, 14 June 1900, NA S.Th 21.2/12.
5. Čhaophrayā Thēwēt to Chulalongkorn, 5 July 1900, NA r5 KS 10/1.
6. Chulalongkorn to Čhaophrayā Thēwēt, 30 June 1900, NA r5 KS 10/1.
7. Čhaophrayā Phātsakǫrawong to Čhaophrayā Thēwēt, 14 March 1902, NA S.Th 21.2/12.
8. Čhaophrayā Thēwēt to Chulalongkorn, 23 February 1901, NA r5 KS 10/1.
9. Nāi Čharoen to Čhaophrayā Thēwēt, 12 January 1901; Chulalongkorn to Čhaophrayā Thēwēt, 25 February 1901, NA r5 KS 10/1.
10. Čhaophrayā Thēwēt to Chulalongkorn, 31 August 1901; Nāi Čharoen to Čhaophrayā Thēwēt, 19 July 1901, NA r5 KS 10/2.
11. Meeting of the Council of Ministers, 5 September 1901, NA r5 KS 10/2.
12. E. Schneegans to Prince Devawongse, 28 December 1901, NA r5 KS 10/2.
13. Chulalongkorn to Čhaophrayā Thēwēt, 22 January 1902, NA r5 KS 10/2.
14. Prince Sommot to Prince Damrong, 29 April 1902, NA r5 KS 10/1.
15. It is interesting to note that just prior to Nāi Čharoen's return from Saigon, the King ordered an investigation into the circumstances in which he had been appointed to the administration in 1900. Čhaophrayā Thēwēt to Čhaophrayā Phātsakǫrawong, 12 March 1902, NA S.Th 21.2/12.
16. The sericulture work of this group will be considered in Chapter 5.
17. Toyama to the Assistant Minister of Agriculture, 11 February 1903, NA r5 KS 10/1.
18. David B. Johnston, 'Rural Society and the Rice Economy in Thailand, 1880–1930', Ph.D. diss., Yale University, 1975, p. 346; *Bangkok Times*, 12 November 1909.
19. Prince Phenphat to Čhaophrayā Thēwēt, 26 October 1904, NA r5 KS 10/1.
20. Prince Phenphat may also have been influenced by his own experience, for his education in Europe had included training in agricultural science. Ministry of Agriculture, *Prawat krasuangkasēt* (*History of the Ministry of Agriculture*), Bangkok, 1957, p. 71.
21. At this time it was frequently argued that the absence of standardized weights and measures in the kingdom made it possible for rural merchants and middlemen, trading in agricultural produce, to defraud cultivators on a substantial scale.
22. Čhaophrayā Thēwēt to Chulalongkorn, 12 November 1904, NA r5 KS 10/1.
23. Chulalongkorn to Čhaophrayā Thēwēt, 26 January 1905, NA r5 KS 10/1.
24. Johnston, op. cit., p. 348.
25. Ministry of Agriculture, op. cit., pp. 72–4; Čhaophrayā Wongsānupraphat, *Prawat krasuangkasētrāthikān* (*History of the Ministry of Agriculture*), Bangkok, 1941, pp. 188–90.
26. Čhaophrayā Wongsānupraphat, op. cit., pp. 245–6.
27. *Bangkok Times*, 12 November 1909.
28. W. A. Graham, *Siam*, 3rd ed., London, 1924, vol. I, p. 355.
29. *Bangkok Times*, 24 August 1912.
30. Tej Bunnag, *The Provincial Administration of Siam 1892–1915: The Ministry*

of the Interior under Prince Damrong Rajanubhab, Kuala Lumpur, 1977, pp. 231–5, 244.
31. Bangkok Times, 24 August 1912.
32. Ibid.
33. Ibid.
34. Čhaophrayā Wongsānupraphat to Chulalongkorn, 11 October 1909, NA r5 KS 10/5.
35. Čhaophrayā Wongsānupraphat to Chulalongkorn, 30 November 1909, NA r5 KS 10/5.
36. Chulalongkorn to Čhaophrayā Wongsānupraphat, 1 December 1909, NA r5 KS 10/5.
37. Ibid.
38. Ibid.
39. See, David K. Wyatt, *The Politics of Reform in Thailand: Education in the Reign of King Chulalongkorn*, New Haven and London, 1969.
40. Čhaophrayā Wongsānupraphat, op. cit., p. 196.
41. Over a decade later W. A. Graham, the Adviser to the Ministry of Lands and Agriculture, was still able to argue that the failure of the provincial officials of the Ministry of Agriculture and of the Ministry of the Interior to create an effective working relationship 'has been responsible for much wasted time and labour'. 'Note on Economic Development and Conservancy', 10 October 1925, NA SB 2.8/7.
42. Johnston, op. cit., pp. 371–3.
43. For details on Čhaophrayā Wongsānupraphat's advocacy, see p. 82.
44. David Feeny, *The Political Economy of Productivity: Thai Agricultural Development, 1880–1975*, Vancouver and London, 1982, pp. 52–3; Johnston, op. cit., p. 363. See also, Dr Yai Suvabhan Sanitwongse, *The Rice of Siam*, Bangkok, 1927, which includes a brief biography of Suvabhan (in Thai) by Prince Damrong. Note should also be made here of the agricultural interests of Mǫm čhao Sithiporn Kridakara, a son of Prince Naret (the eldest of King Chulalongkorn's half-brothers). Educated at Harrow, Mǫm čhao Sithiporn enjoyed a most distinguished career in the Siamese administration of the early twentieth century, rising to the position of Director-General of the Opium Department. But in 1920, at the age of 37, he resigned from the government service to administer a farm at Bangbert, some 400 kilometres south of Bangkok. Thereafter he devoted the remainder of his life (although he was imprisoned between 1933 and 1944 for alleged involvement in the Bowaradej rebellion) to agricultural experimentation and to the advance of the kingdom's agrarian interests. See, Sithiporn Kridakara, *Some Aspects of Rice Farming in Siam*, Bangkok, 1970; Benjamin A. Batson, [Review Article], *M. C. Sithiporn Kridakara Memorial Volume*, *Journal of the Siam Society*, vol. 61, pt. 1 (January 1973), pp. 302–9.
45. See for example, Ministry of Agriculture, op. cit., p. 88; Čhaophrayā Wongsānupraphat, 'Memorandum on our Domestic Economy' (in English), 7 December 1910, NA r6 KS 1/4.
46. See pp. 26 and 28.
47. It would appear that this practice was also common in Indo-China. See Charles Robequain, *The Economic Development of French Indo-China*, London, 1944, pp. 309–10.
48. Ministry of Agriculture, op. cit., pp. 88–90.
49. Feeny, *The Political Economy of Productivity*, pp. 54–5.

50. Ibid., p. 55.
51. Ibid., p. 58.
52. It might be added that the Rangsit district was not always the administration's favoured site for an experimental farm. As was noted earlier (p. 65), Prince Phenphat planned farms in Rātburī and in Nakhǫn Pathom; and, according to one source, in the year prior to the establishment of the Rangsit farm, an experimental agricultural station had been created in Phitsanulōk province. See Ministry of Agriculture, op. cit., p. 121.
53. Johnston, op. cit., pp. 363–6.
54. Ibid., pp. 366–71.
55. Ministry of Agriculture, op. cit., p. 89.
56. Ibid., p. 143.
57. Feeny, *The Political Economy of Productivity*, p. 55.
58. Ibid., p. 140.
59. V. D. Wickizer and M. K. Bennett, *The Rice Economy of Monsoon Asia*, Stanford, 1941, p. 238.
60. Robequain, op. cit., pp. 228–9.
61. Cheng Siok-Hwa, *The Rice Industry of Burma 1852–1940*, Kuala Lumpur, 1968, pp. 39–40.
62. Glenn Anthony May, *Social Engineering in the Philippines: The Aims, Execution, and Impact of American Colonial Policy, 1900–1913*, Westport, Conn., 1980, p. 140.
63. Lewis E. Gleeck Jr., *American Institutions in the Philippines (1898–1941)*, Manila, 1976, p. 232.
64. Michael Adas, *The Burma Delta: Economic Development and Social Change on an Asian Rice Frontier, 1852–1941*, Madison, Wisconsin, 1974, p. 131.
65. Cheng, op. cit., p. 28.
66. Yujiro Hayami, *A Century of Agricultural Growth in Japan: Its Relevance to Asian Development*, Tokyo, 1975, p. 228.
67. The following is drawn from: Hayami, op. cit., specifically pp. 52–3, 55; Shigeru Ishikawa, *Essays on Technology, Employment and Institutions in Economic Development: Comparative Asian Experience*, Tokyo, 1981, specifically p. 158.
68. Ishikawa, op. cit., p. 158.
69. These reforms included notably the Land Tax Revision which replaced the feudal tax in kind (levied in proportion to quantities harvested) by a land tax in cash (based on the value of land), and an associated nation-wide cadastral survey (undertaken in the years 1873–81); and two measures introduced at the beginning of the 1870s which removed an existing prohibition on the sale and mortgage of farmland, and which left farmers free to determine land utilization and crop patterns. Hayami, op. cit., pp. 46–8.
70. Hayami, op. cit., p. 60. It should be added that by the end of the Meiji era this backlog of indigenous technological potential had been exhausted, and consequently more basic research (involving principally artificial cross-breeding programmes) was then required to sustain the growth in agricultural productivity. Ibid., pp. 61, 64–6.
71. Ibid., p. 56.
72. Ibid., p. 45.
73. Ibid., p. 207.
74. Prince Damrong to Chulalongkorn, 18 May 1906, NA r5 KS 3.1/11.

Johnston has calculated that in the seven years from 1905/6, the population of Thanyaburī district (which may be considered coterminous with the Rangsit concession) fell by some 35 per cent (Johnston, op. cit., pp. 312–14).

75. 'Rāingān kamakān truattaisuan ru'ang nā naithungluangrangsit' ('Report of the Committee to investigate [conditions] in the Rangsit area'), 4 May 1906, NA r5 KS 3.1/11. Part of this report appears in an English translation in Chatthip Nartsupha and Suthy Prasartset (eds.), *The Political Economy of Siam 1851–1910*, Bangkok, 1981, pp. 429–34.

76. Ibid. This is an amended version of the translation in Chatthip and Suthy.

77. Ibid.

78. Suvabhan Sanitwongse to Prince Damrong, 10 May 1906, NA r5 KS 3.1/11.

79. Chulalongkorn to Čhaophrayā Thēwēt, 22 May 1906, NA r5 KS 3.1/11.

80. *Bangkok Times*, 15 September 1909.

81. *Bangkok Times*, 20 September 1909.

82. *Bangkok Times*, 25 September 1909.

83. *Bangkok Times*, 24 September 1909.

84. *Bangkok Times*, 25 September 1909, 29 September 1909.

85. *Bangkok Times*, 24 September 1909, 1 October 1909.

86. *Bangkok Times*, 4 October 1909.

87. *Bangkok Times*, 7 January 1910.

88. *Bangkok Times*, 28 September 1909.

89. *Bangkok Times*, 23 September 1909.

90. *Bangkok Times*, 25 September 1909, 4 October 1909.

91. Čhaophrayā Wongsānupraphat to Chulalongkorn, 18 August 1909, NA r5 KS 3.1/12.

92. Ibid.

93. In fact these precise points had been made by W. J. Archer, 'Memorandum on the Establishment of the Agricultural Bank of Egypt', 6 January 1909, NA r5 KS 10/4.

94. A correspondent to the *Bangkok Times* (28 September 1909) argued from the experience of Egypt that an agricultural bank would not succeed in Siam until adequate irrigation works were constructed.

95. Čhaophrayā Wongsānupraphat to Chulalongkorn, 18 August 1909, NA r5 KS 3.1/12.

96. Meeting of the Council of Ministers, 20 August 1909, NA r5 KS 3.1/12.

97. Chulalongkorn to Čhaophrayā Wongsānupraphat, 21 August 1909, NA r5 KS 3.1/12.

98. Čhaophrayā Wongsānupraphat to Chulalongkorn, 11 October 1909, NA r5 KS 10/5.

99. Chulalongkorn to Čhaophrayā Wongsānupraphat, 13 October 1909, NA r5 KS 10/5.

100. Čhaophrayā Wongsānupraphat to Chulalongkorn, 23 December 1909, NA r5 KS 3.1/12.

101. 'Rāingān prachum chāwnā nai thungkhlǭngrangsit' ('Report on a meeting of Rangsit landowners'), 10 December 1909. Enclosed in, Čhaophrayā Wongsānupraphat to Chulalongkorn, 23 December 1909, NA r5 KS 3.1/12. See also, *Bangkok Times*, 6 January 1910.

102. Čhaophrayā Wongsānupraphat to Chulalongkorn, 23 December 1909, NA r5 KS 3.1/12.

103. Chulalongkorn to Čhaophrayā Wongsānupraphat, 25 December 1909, NA r5 KS 3.1/12.
104. Petition to Čhaophrayā Wongsānupraphat, 24 January 1910. Enclosed in, Čhaophrayā Wongsānupraphat to Chulalongkorn, 2 March 1910, NA r5 KS 3.1/12.
105. Čhaophrayā Wongsānupraphat to Chulalongkorn, 2 March 1910, NA r5 KS 3.1/12.
106. Chulalongkorn to Čhaophrayā Wongsānupraphat, 6 March 1910, NA r5 KS 3.1/12.
107. Čhaophrayā Wongsānupraphat to Vajiravudh, 9 December 1910, NA r6 KS 1/5.
108. The official history of the Ministry of Agriculture indicates that this was due to the administrative disruptions which attended the death of King Chulalongkorn (in October 1910) and the transfer of Čhaophrayā Wongsānupraphat to another ministry early in the new reign. (Ministry of Agriculture, *Prawat krasuangkasēt*, p. 92.)
109. Čhaophrayā Wongsānupraphat, 'Memorandum on our Domestic Economy', 7 December 1910, NA r6 KS 1/4. (In English.) Part of this document is reproduced in, Chatthip Nartsupha, Suthy Prasartset and Montri Chenvidyakarn (eds.), *The Political Economy of Siam 1910-1932*, Bangkok, 1981, pp. 219-22.
110. Ibid.
111. Ibid.
112. Ibid.
113. Ibid. Čhaophrayā Wongsānupraphat's poor opinion of the character of the Siamese cultivator contains, perhaps surprisingly, an echo of widely-held European attitudes. For example, writing in the mid-1920s, W. A. Graham, who had served the Siamese administration in a number of capacities since the early years of the century, noted, 'The peasant has few personal wants and almost no ambitions. His moral fibre is weak and he has always been an easy victim to those who prey upon the weak and ignorant.' W. A. Graham, 'Note on Economic Development and Conservancy', 10 October 1925, NA SB 2.8/7.
114. Čhaophrayā Wongsānupraphat, 'Memorandum on our Domestic Economy', 7 December 1910, NA r6 KS 1/4.
115. Ibid.
116. It should be added that Johnston, op. cit., chapter vii, provides a valuable account of the course, immediate impact, and the long-term significance of the recession.
117. David F. Holm, 'The Role of the State Railways in Thai History, 1892-1932', Ph.D. diss., Yale University, 1977, p. 159. Presumably 'ignorant' is used here in the sense of 'gauche'.
118. See, for example, *Bangkok Times*, 1 November, 16 November 1915; W. A. Graham, 'Note on Economic Development and Conservancy', 10 October 1925, NA SB 2.8/7.

3
The Enclaves: Tin and Teak

THE production of tin and teak for export occupied a fundamentally different position in the economy of late nineteenth- and early twentieth-century Siam from that of the cultivation for export of rice. The latter was, of course, by far the most valuable of the kingdom's exports, accounting for around 70 per cent of total export earnings during the opening two decades of the twentieth century compared with around 10 per cent each for tin and teak.[1] But more importantly, whereas the cultivation of rice was undertaken almost exclusively by Siamese, the felling of teak and the extraction of tin was overwhelmingly dominated by immigrant Asians and Europeans; and whilst rice for export was cultivated almost exclusively in the Central Plain of Siam, teak and tin were concentrated in the northern and southern extremities of the kingdom respectively. A further contrast might be made: that whereas tin is a wasting asset and teak stands require strict conservation, there is no comparable restriction involved in the cultivation of rice. For these reasons the production of tin and teak for export confronted the Siamese administration of this period with problems of a radically different nature from those it faced with respect to the rice-export sector.

At the same time, there was also at least one crucial difference between the tin and teak sectors themselves which had important implications for government policy and administration in late nineteenth- and early twentieth-century Siam. In the case of teak, until the mid-1880s exports were slight. However, the temporary closing of the teak forests of Upper Burma at the conclusion of the Third Anglo-Burmese War of 1885 brought the major European timber companies into northern Siam, and consequently a major expansion of Siamese teak exports until the 1900s. In contrast, the export of tin from the southern peninsular provinces of the kingdom expanded significantly from the 1850s until the mid-1870s, at which point Siam's production virtually matched that of the Malay States to the south.[2] However, from the end of that decade tin production in Siam fell back, whilst that in Malaya underwent the dramatic expansion which was to make the British territory the world's dominant tin producer by the end of the century. In short, during this period the principal concern of the Siamese administration with respect to the extraction of teak was to regulate and constrain a vigorous foreign interest; in contrast, with tin it was faced with the decision whether, and in what manner, to *attract* capital, labour, and entrepreneurship into the sector to secure a more rapid exploitation of the resource. This chapter will first consider that last problem.

I

In the first half of the 1890s average annual tin output in Siam was around 4,250 tons, compared with some 6,450 tons in the mid-1870s;[3] in the same period Malayan production had soared to an annual average of 36,300 tons.[4] The depressed condition of the southern industry was closely considered by the Siamese administration from the early 1890s. Three points were strongly emphasized. First, that communications in the tin districts of the peninsula were extremely poor. When H. Warington Smyth, Director of the Department of Mines, visited the principal tin district, the island of Phūket, in the mid-1890s he was 'at once struck by the abominable state of the roads', adding that 'the indifference of the Government to the condition of the roads' imposed a considerable burden on the tin-miners, particularly those working deposits some distance from the main town.[5] With regard to *monthon* Phūket as a whole, in 1908 a British consular official reported that 'the country is almost entirely roadless, except in the immediate vicinity of one or two towns'.[6] And in the mid-1890s, the King's attention was frequently drawn to the almost entire absence of tracks and roads in Phūket, and the consequent restriction on tin extraction in the area.[7] But it was not only land communications which were seriously inadequate. By this period Phūket harbour, the principal outlet for Siam's tin, had become so silted by the tailings from local mines that sea-going vessels were forced to anchor over a mile from the landing stage,[8] cargoes then being loaded and unloaded by lighter. In 1906, the government reached an agreement with a Captain Edward T. Miles, a Tasmanian, as part of which the latter was to dredge an entrance to the harbour and a dock at Phūket.[9] But it soon emerged that the harbour improvements were to be more difficult and costly than had been anticipated, and consequently by the end of the decade the Siamese Government, in return for a cash settlement, released Miles' company from that project.[10] In brief, in the late nineteenth and early twentieth centuries peninsular Siam did not appear to have possessed either the port facilities or the internal communications that would permit a major expansion in tin excavation and export.

Second, attention was frequently drawn in the 1890s to a serious shortage of Chinese coolie labour in the tin districts. In the 1860s and 1870s there had been a substantial influx of Chinese labour into Phūket,[11] but from the late 1880s it was found that a major proportion of the coolie migrants recruited in China was leaving ship on arrival at either Singapore or Penang, rather than proceeding to the Siamese port.[12] In addition, an increased demand for manual labour in Central Siam also from the late 1880s appears to have diverted Chinese migrants away from the tin districts of the peninsula.[13] The peninsular districts not only failed to attract fresh labour in this period but also lost a very substantial proportion of their existing work-force, principally to British-administered Perak and Selangor to the south.[14] In the mid-1880s there had been an estimated 50,000 Chinese tin-miners in Phūket; a decade later there were only 11,000.

The drain of Chinese labour from Phūket at the close of the nineteenth century reflected not only the very rapid expansion of tin production in Malaya in that period (and to a lesser extent the increase in opportunities for migrant labour in Central Siam) but also, in the view of a number of observers, the severely discouraging conditions under which Chinese laboured in the Siamese-administered portions of the peninsula. For example, in early 1893 Phrayā Thipkōsā, the commissioner in Phūket, noted in a letter to Bangkok that tin labourers in the region were not assured an agreed wage but were paid in accordance to the returns on the mine in which they worked.[15] If a mine lost heavily, its workers would receive no wage payment. Three years later, the King noted that in British Malaya the contractual conditions of work for Chinese tin-miners were regulated and inspected by the administration, but that no comparable protection for labour was to be found in the Siamese tin districts.[16] For in those districts the governors were themselves mine-owners, and thus were interested only in extracting the maximum profit from their mining operations, not with the contract conditions for Chinese labour. European officials in the

Siamese administration frequently argued, in more general terms, that the coolie was harshly exploited by the local élite. Writing to the acting Minister of Agriculture in January 1895, Walter de Müller, Director of the Department of Mines, noted that Phūket 'is in the hands of a few rich Chinese men, who try to make as much money as they can by grinding down their workmen'.[17] He added that the Chinese workers were also being driven from Phūket by the high cost of essential commodities, including rice, caused by the imposition of heavy import duties. The poor quality of the opium supplied in Phūket added to the labourers' discontent.[18]

Third, it was strongly argued that the prospects for tin mining in the Siamese peninsular states were being 'ruined by ... exorbitant taxes'.[19] In the mid-1890s it was reported that the tin royalty charged in Phūket was equivalent to over 16 per cent.[20] For comparison, the average annual rate of tin duties levied in Perak and Selangor in that decade ranged from 8.4 per cent to 13.4 per cent, although principally in the upper part of that range.[21] But, in addition to the tin royalty (and other imposts stipulated by the central government), mining in southern Siam was also required to bear a considerable number of charges imposed on the initiative of the local administration. When H. G. Scott, then Director of the Department of Mines, toured Phūket in 1900, he discovered that some 8 taxes were being imposed on tin production in the district, in addition to those specified in government mining regulations.[22] Finally, reference should be made to a calculation by H. Warington Smyth in the mid-1890s that 'taking all the farms and taxes [in Phūket] into consideration, the Government took forty per cent of the earnings of every coolie in the place'.[23]

If the decline in the peninsular tin industry in the early 1890s was most frequently attributed to heavy taxation, inadequate communications, and a serious shortage of Chinese labour, attention was also drawn to a number of other considerations. Prominent here was the argument that recent changes in government financial administration in the south had severely reduced the capital available for tin mining. Both the King and Prince Damrong noted that the earlier development of the industry had been financed principally by the local governors, since the form of provincial administration then in force there permitted them to retain the revenues collected under their jurisdiction.[24] But, from the mid-1870s the central government had insisted, with increasing effectiveness, that by far the major part of those revenues be transmitted to Bangkok, and this had left the governors unable to maintain their earlier investment in the industry. Mining entrepreneurs had sought funds in Penang, but capital there was flowing mainly into the Malay States under British administration. A further point, made by H. Warington Smyth in 1896, was that although there was 'more tin still unworked [in Siam] than ever came out in the [British] Malay States', nearly all the rich alluvial tin deposits (particularly in Phūket) were being worked out and miners now faced the increased difficulty and expense of exploiting the poorer and more remote seams.[25] Finally, attention was frequently drawn in this period to the absence in Siam of effective legal and administrative structures to regulate mining operations in the kingdom. Thus, it was common practice for miners to work an area without first securing permission from the administration; there were no established procedures for setting the boundaries of mining concessions.[26] Neither were there regulations to prevent miners from diverting water channels (which were essential for mining operations) to the severe disadvantage of others.[27] Conflict over mine boundaries and water rights was therefore endemic.[28] In brief, there was no legal or administrative security in the Siamese tin districts for the established mining entrepreneur.

Perhaps the most pointed observations on the depressed condition of the Siamese tin industry at the end of the nineteenth century came from the King himself, in a letter to Čhaophrayā Thēwēt in late 1896.[29] After maintaining (erroneously) that the tin resources of Siam exceeded those of the Malay States, Chulalongkorn explained the far more rapid development of the Malayan industry essentially in terms of the vigorous commitment of the British administration to its progress. Cart-tracks and railways had been constructed to facilitate haulage to and from the mines; Chinese coolie immigration had been actively assisted; the tin royalty had been reduced in order to encourage investment in the Malayan tin districts. Was a comparable commitment to be seen on the Siamese side of the border?

In summary, in the early 1890s the Bangkok administration was clearly aware of the backward condition of the tin industry in the peninsular provinces, and was forming firm ideas as to the causes of that backwardness. Perhaps more importantly, it also had a keen appreciation of the progress then being achieved in the Malay States under British rule, and thus presumably an awareness of the potential for expansion within the Siamese tin industry. How was Bangkok to respond?

II

The central government laid the foundation for effective regulation of the southern tin industry by the establishment in Bangkok, in January 1892, of a Department of Mines.[30] The Department, which came under the Ministry of Agriculture, had as Director, Walter de Müller, a German, and as his Deputy, H. Warington Smyth, an Englishman. An important early concern of the Department was to secure detailed knowledge of the mining concessions then being worked in the peninsula, and to this end inspection tours were undertaken in the south, notably Phūket, in this period.[31] These, however, do not appear to have been particularly successful, for frequently the local administrative and commercial élite (who, of course, had major interests in the industry) withheld assistance from the Bangkok officials.[32] In fact, effective regulation of tin mining in the peninsula by the central government had to wait until those provinces had been brought within the reformed system of provincial administration then being advanced by Prince Damrong. It is therefore important to note that in precisely the same year as the first superintendent commissioner was appointed to the newly-created *monthon* Phūket, that is 1898,[33] a provincial office of the Department of Mines was established in the *monthon* capital, to which was appointed a mines commissioner with responsibility for issuing concessions and inspecting leases throughout the southern provinces.[34] A second mines commissioner was appointed (to *monthon* Nakhǭn Sīthamarāt) the following year, and by 1915 there were six such officials posted in the south.[35] These appointments, in alliance with the reform of provincial administration in the peninsula, undoubtedly secured for the central government considerable authority over mining activity in the south, although whether Bangkok's authority in this period could compare with that exercised by the British administration in Malaya (where the major centres of government were situated directly in the principal tin-yielding areas) is less clear.

The second major task of the Department of Mines in its early years was to draft mining legislation for the kingdom. Drawing heavily on legislation in force in British Malaya in particular,[36] by early 1895 the Department had prepared a draft Mining Act, and in March that year the Council of Ministers appointed a special committee, which included Phrayā Surasakmontrī (the Minister of Agriculture) as chairman, Prince Damrong and H. Warington Smyth (by then the Director of the Department of Mines), to review it.[37] The committee's most important discussion concerned the level of taxation to be imposed on tin production in the kingdom. Clearly familiar with the relatively moderate rates in force in British Malaya and obviously disturbed by the diverging fortunes of the Siamese and Malayan industries, it proposed that the maximum rate of tin royalty be 10 per cent, the annual land rent be fixed at 4 baht a hectare, and the tin stamp tax (*phāsī tītrā dībuk*) be abolished. The committee was confident that the substantially reduced burden of taxation implied by these rates would attract a greatly increased number of miners into the Siamese tin districts. When the draft legislation came back to the Council of Ministers in September 1896, the Council insisted on two amendments: that the maximum rate of tin royalty be raised to 17 per cent; and that a clause be inserted requiring applicants for mining leases to provide a deposit of up to 10,000 baht to demonstrate good faith.[38] These amendments raised a critical question as to the direction of government mining policy. On one side it was argued (notably by the Director of the Department of Mines) that the high royalty in particular would condemn the southern tin industry to continued stagnation:[39] yet other members of the administration (notably the Minister of Finance, Prince Mahit) were apparently less disturbed by the continued stagnation in tin production than by the prospect of an initial fall in the revenue from tin if the lower rate were imposed.[40] After considerable discussion within the administration, the King instructed that the original proposal (that the maximum rate be 10 per cent) be restored.[41]

The progress of the mining legislation was then long delayed by the need to have it sanctioned by the treaty powers.[42] Without that approval the legislation could not apply to foreign subjects; and of course the tin-miners at work in Siam were overwhelmingly non-Siamese. The Mining Act was finally promulgated in September 1901, its 84 clauses providing detailed regulation for the granting and working of mining concessions in the kingdom.[43] It was implemented in the principal tin districts of the peninsula from the middle of the following year. This was a major administrative task, for it required the provincial officials of the Department of Mines to investigate and survey all the existing tin workings in their region and, if appropriate, to confirm the rights of miners to their concession within the provisions of the new legislation. Given the unregulated manner in which mining rights had earlier been claimed and worked, this process was clearly an extremely arduous and time-consuming one, presumably the more so in those districts (away from Phūket island) where communications were particularly poor, and effective administration remained less secure. In thus drawing out the difficulties and delays which confronted the Bangkok administration in this period as it sought to provide a secure legal structure for the tin industry in the peninsula, an instructive comparison may again be made with the conditions within which the neighbouring British administration worked. For in framing mining legislation the latter could call upon the experience of an empire-wide bureaucracy; enactment was not impeded by treaty restrictions; and most importantly, implementation was facilitated by the relative effectiveness of central administration in the tin districts themselves. These were important advantages for the colonial authority.

III

The establishment of a mining administration in Bangkok and in the peninsular provinces, and the introduction of legislation to regulate mining activity in the kingdom, coincided with a modest revival in Siam's tin production. The average annual output of tin from Siam in 1890–4 had been 4,236 tons; in 1910–14 it was 5,776 tons.[44] But the early twentieth century also saw important changes in the structure of the Siamese tin industry. Perhaps the most radical was the introduction in late 1907 of the bucket dredge, brought to Siam by an Australian concern, the Tongkah Harbour Tin Dredging Company.[45] By the beginning of the 1920s there were 13 dredges at work in the peninsula, 5 of which were owned by the pioneer Australian company. In 1930 there were 38 dredges in operation or in course of erection in Siam.[46] The superiority of the bucket dredge over the labour-intensive methods long employed in the peninsula lay primarily in its ability to work profitably, land where the tin content was low. Indeed, it was possible for the bucket dredge to work successfully, concessions which had already been exploited by labour gangs. As nearly all the rich alluvial deposits (particularly in Phūket) were being worked out by the closing years of the nineteenth century,[47] the proportion of tin output secured by dredge increased steadily from the late 1900s and indeed exceeded non-dredge production at the end of the 1920s, although the absolute level of non-dredge output also rose in this period.[48] But critically the Chinese mining entrepreneurs in the peninsula did not adopt the bucket dredge, and consequently the advance of the new technology implied the relative eclipse of the Chinese interest in the industry.[49] There were important implications here for the Bangkok administration. At the end of the nineteenth century the central government had sought 'to blunt the forward drive of British economic incursions' in the peninsula in part by appointing local Chinese capitalists, with major interests in tin, to important political positions in the region.[50] By the end of the 1900s, it was clear that the changing technological demands of tin mining in the peninsula had undermined that strategy, and that within a short period Western capital would dominate the industry.

The increasing Western interest in the extraction of Siamese tin was accompanied by the extension of Western control over the smelting of the kingdom's ore. Until the end of the nineteenth century, the ore from Chinese mines had been smelted locally in 'little iron-bound mud blast-furnaces', fuelled by charcoal taken from nearby forests.[51] Comparable techniques had been almost exclusively employed in the British territory to the south until the end of the 1880s when the newly-established Straits Trading Company, challenging the Chinese smelting monopoly, erected a modern smelter on Pulau Brani, a small island just off Singapore.[52] A second Straits Trading smelter, at Butterworth (on the mainland, directly opposite Penang), began operation in 1901. Towards the end of that decade a third modern smelter was constructed, in Penang. Here the initial investment was made primarily by Chinese interests, but in 1911 the concern, the Eastern Smelting Company, was bought out by British capital.[53] Straits Trading and Eastern Smelting rapidly commanded increasing supplies of Siamese ore. In fact, as early as 1890 the former company asked the Siamese administration for a monopoly on the purchase and export of tin ore from the peninsular provinces for a period of 10 years.[54] The request was refused—but it soon became apparent that the company required no such agreement. The presence of large-scale modern smelters within a short shipping distance of the Siamese tin provinces inevitably led to an increasing proportion of Siam's production being exported in the form of ore for smelting in the British territory and the collapse of local Chinese smelting.[55] In 1909–10 well over 60 per cent of the tin exported from *monthon* Phūket was shipped as ore;[56] at the beginning of the 1920s the amount of tin-ore smelted locally by Chinese smelters was 'almost negligible',[57] and by 1930 all tin exports from Siam were in the form of ore, over 90 per cent being shipped directly to British Malaya.[58] Inevitably the monopsonist position of the British smelting companies in Malaya could only cause further concern in Bangkok.

Although the scale of Western investment in the Siamese tin mining industry in the early twentieth century and the expansion in production which resulted in part from it were very modest,[59] the Bangkok administration viewed these changes with considerable apprehension. The minister with principal concern in this respect was Prince Damrong, not simply because the Ministry of the Interior was responsible for administration in the peninsula but more particularly because from 1896 to 1909 the Department of Mines was directly under its authority.[60] On occasion the Minister was prepared to offer some direct, if limited, encouragement to increased tin production, as for example in late 1901 when he supported a call from the Director of the Department of Mines that the government temporarily waive the royalty on tin from Ranǭng in order to attract miners into that district.[61] But far more frequently Prince Damrong was concerned that the more rapid growth of the peninsular industry would almost inevitably bring with it damaging repercussions.

An important principle of the Minister of the Interior in this period was that large-scale mining should take place only in those districts where there was strong local administration. When, in late 1894, de Müller claimed that the mining industry in the peninsula was being 'ruined by ... exorbitant taxes' while the government did nothing to provide adequate roads in the tin districts,[62] Prince Damrong replied that Bangkok's primary concern in the southern provinces must be to establish the *Thēsāphibān* system of provincial administration, and only then would it be in a position to encourage the industry.[63] Strong provincial administration was essential not because it would provide a secure (and thus attractive) environment for mining capital but rather because it would ensure that the disruptive elements commonly associated with mining could be firmly controlled. Prominent here were the riots and gang-warfare which frequently erupted among the Chinese coolie population in the tin districts. Writing in the mid-1890s, Smyth referred to a recent occasion on which nearly 20,000 riotous Chinese threatened Phūket town.[64] That mob had been dispersed by troops under Commodore de Richelieu, a Danish officer in the Siamese Navy, but it remains doubtful whether the government forces in the area in this period—a company of men from Bangkok, a single gunboat, and a force of some 60 police—could have suppressed a concerted uprising. Throughout the late 1890s and 1900s Bangkok received disturbing reports of violent, frequently fatal clashes between Chinese miners in the peninsula,[65] and in March 1906 the King admitted to Prince Damrong that he was constantly worried that the administration in the south was still insufficiently secure to be able to bring serious disorder under control.[66] Undoubtedly, Chulalongkorn's underlying fear was that instability in the peninsular provinces would almost inevitably attract the interest of the British administration to the south, and that as Western capital began to flow into the Siamese tin districts in the later 1900s, that interest could so easily become more interventionist.

But if there was serious concern over the presence of a large and unruly Chinese mining population in the south, there were comparable misgivings at the prospect of substantial numbers of Europeans moving into the peninsula in search of mining concessions. When, in late 1898, concern was expressed in the administration that the tardiness of the treaty powers in sanctioning the mining legislation was seriously delaying its enactment,[67] Prince Damrong advised that the government simply keep quiet for the time being—for in fact he 'did not wish to open up the peninsular provinces to foreigners who would crowd in, demanding concessions'.[68] The Minister later made it clear that his fears here centred less on *bona fide* Western mining entrepreneurs such as Captain Miles of the Tongkah Harbour Tin Dredging Company than on the far greater number of individuals who sought leases solely for speculative purposes or more usually in order to establish flimsy companies which would attract funds from naïve investors.[69] Prince Damrong's strong dislike of this 'objectionable type of adventurer'[70] derived not from their lack of interest in actual mining but rather from the fact that in his view their shady activities seriously damaged the international reputation of the kingdom.

In fact, during this period Prince Damrong's wish to exclude Western concession seekers from the peninsula was achieved in part by a secret convention concluded between Siam and Britain in

April 1897.[71] Under the convention Siam undertook not to cede to a third power any part of her territory south of Bāngtaphan. For Britain the agreement secured her strategic position in the Straits of Malacca, while for Siam it indicated British acknowledgement of her sovereignty over the states of Kedah, Kelantan, Trengganu, and Perlis. In drafting the convention, it had been recognized that were the subjects of a third power allowed to acquire land leases or trading rights in the peninsula, those concessions might be used to advance that power's political influence in the area, so undermining the principal aim of the agreement. Consequently, the final article of the convention required Siam not to grant 'any special privilege or advantage whether as regards land or trade' in the peninsula to the subject of a third power 'without the written consent of the British Government'. In practice, all applications from the subjects of a third power for land or trade concessions in this region of the kingdom, including of course applications for prospecting licences and mining leases, had to be referred to the British Minister in Bangkok. The latter was always likely to reject them—indeed during the life-time of the convention the large commercial concessions granted by the Siamese Government in the peninsular provinces were worked exclusively by British companies or British capital. In other words, although the initial object of the 1897 secret convention was to prevent any third power establishing a strong political influence in the peninsula, it became, largely unforeseen, primarily a means by which Britain excluded non-British capital from the southern provinces of Siam. If the administration in Bangkok was pleased to have this barrier erected against an influx of Western concession seekers into the peninsula, it must be said that this advantage was outweighed by the serious difficulty it faced in explaining its delay in dealing with, and then frequent rejection of, commercial applications from non-British subjects (notably Germans), without making reference to the secret agreement with Britain. Moreover, with each rejection Siam was open to the charge that she was denying foreign subjects rights granted them by the treaties concluded in the 1850s and 1860s. And, of course, the convention did not restrict the advance of British concession holders in the southern provinces. In March 1909, as Siam and Britain reached agreement over the transfer of the four northern Malay States to British authority, the secret convention of 1897 was cancelled.

However, it is crucial to note that within a few months of the removal of this barrier, Prince Damrong was seeking to impose a further, if less severe, restriction. His proposal arose from the government's decision to proceed with construction of the peninsular railway, the project to be financed by means of a loan from the Federated Malay States agreed in March 1909.[72] The railway was to be laid on the east side of the peninsula, in part on the insistence of the Singapore business community which wanted that port rather than Penang to benefit from the new line, and would therefore pass some considerable distance from the principal tin districts of *monthon* Phūket. However, with the line constructed down the east coast, the commercial potential of the tin deposits there, which until that time had attracted relatively little investment, was almost certain to be considerably enhanced. It was therefore Prince Damrong's fear that with the decision to proceed with the railway, those districts would now be flooded with concession hunters who had no intention of undertaking mining themselves, but, rather, would sell their rights as their value rose.[73] Consequently, in August 1909 he proposed that for an appropriate period no further prospecting licences or mining leases be issued for *monthon* Chumphǫn, Nakhǫn Sīthamarāt, and Pattānī. The proposal was approved by the Council of Ministers at the end of that month.[74]

Prince Damrong's hostility towards the great majority of Western mining interests seeking concessions in Siam frequently brought him into conflict with the European officials in the Department of Mines. In a memorandum written in August 1909, the Minister outlined their opposing approaches to mining administration:

The Europeans in the Department of Mines believe that the establishment of the Department and the enactment of mining legislation has been carried out simply in order to open up every district in the kingdom where there are mineral deposits to people of every nationality and tongue.... It is my opinion that ... [mining] should be permitted only in those districts where effective local administration had been established and where communications were good.[75]

A recent manifestation of this conflict of views, Prince Damrong indicated, had occurred over his proposal for a temporary prohibition on the issue of prospecting licences in the east coast provinces of the peninsula, the European officials arguing that the prohibition would bring protests from foreign powers that Siam was in breach of the treaties concluded in the mid-nineteenth century. More generally, the Europeans had challenged Prince Damrong's view that the shady activities of Western concession hunters in Siam would damage the international reputation of the kingdom. The Minister was assured that sharp practice was common in mining throughout the world, and that when this took place the fault lay with the mining promoters and those investors who had blindly believed them, not with the government which had granted prospecting or mining rights.[76] In brief, Prince Damrong had reason to conclude that the Europeans in the Department of Mines were more concerned to encourage an expansion of Western mining interests in the kingdom, even if this brought in a large fraudulent element, than to protect what he held to be the interests of Siam. His suspicions in this respect may have been strengthened by the fact that during the 1890s there was a frequent turnover in European directors of that Department (between 1892 and 1898 four Europeans held that post),[77] and more particularly by the fact that when the last of those appointments, H. G. Scott, left the government's service in 1907, he did so in order to undertake mining on his own account in Phūket.[78] In October of that year Prince Damrong argued that as tin mining in the peninsula had acquired serious political and administrative associations, the two most senior positions in the Department of Mines should be occupied by Siamese officials, and Europeans should be employed only as inspectors or in a technical capacity.[79] When Scott resigned, a Siamese director was appointed.[80]

Prince Damrong's cautious and restrictive approach to mining administration was opposed not only by European officials in the Department of Mines but also by his nephew, Crown Prince Vajiravudh. After a visit to Phūket towards the end of the 1900s the latter argued that while the Department of Mines had been part of the Ministry of the Interior it had confined itself to bureaucratic work (essentially the issue of mining licences and leases) and had made no attempt itself to prospect for new mineral deposits in order to enhance the kingdom's wealth.[81] It was at the instigation of the Crown Prince that in July 1909 the Department was removed from Prince Damrong's Ministry and returned to the Ministry of Agriculture, in the hope that being part of a small but growing ministry it would have a greater opportunity to develop a more active role.

The Department's new minister, Chaophrayā Wongsānupraphat,

certainly took a more favourable (or perhaps more resigned) view of the effect on Siam's mining industry of substantial foreign investment. In a memorandum written in October 1911, he first accepted that the construction of the southern railway would indeed encourage foreign investment in mining (as well as in agriculture and forestry) in the peninsular provinces.[82] But whereas Prince Damrong had sought to impede that capital inflow, Čhaophrayā Wongsānupraphat argued that as long as the Siamese themselves had neither the capital resources nor the ability required to undertake such major commercial projects, no effective defence could be erected against it. He compared the behaviour of foreign capital to the action of a river in flood. Although dikes and dams might be built to protect an area from flooding, water (foreign capital) would always create a passage through which it could flow. More importantly, Čhaophrayā Wongsānupraphat established that a major inflow of capital, rather than being inevitably harmful for the recipient economy, could be harnessed to its advantage. Indeed he argued that countries like Siam (and the United States) which had extensive lands but limited population and capital resources depended upon foreign investment to open up their land to production, and so increase trade and government revenue. In brief, Čhaophrayā Wongsānupraphat would encourage foreign investment in the kingdom, in mining and agriculture alike. This analysis produced one immediate practical proposal. Also in October 1911, the Minister of Agriculture proposed that the prohibition on the issue of prospecting licences and mining leases for the east coast provinces of the peninsula, imposed at the instigation of Prince Damrong in 1909, be lifted on the grounds that the southern railway would not be profitable until tin production in the area rose.[83] The prohibition was removed early the following year.

This appears to have been the last substantial initiative made by the Bangkok administration with respect to the peninsular tin industry until the beginning of the 1930s. In 1918, King Vajiravudh received a petition from a businessman in Ranǫng, Koh Tiew Lim, pointing out that as Phūket's tin output was now smelted exclusively by two British companies in Penang, foreign interests were securing a major income at the expense of the producer in Siam; and he proposed that the government encourage the establishment of smelting facilities in Phūket by lowering the tax on refined tin while maintaining the rate imposed on tin ore.[84] However, although Chulalongkorn had shown interest in the construction of a modern smelter on the island as far back as the early 1890s,[85] and on this occasion Koh Tiew Lim was invited to Bangkok to discuss his proposal,[86] nothing came of the matter.[87] The following year, 1919, a new Mining Act was introduced, but this legislation was intended primarily to correct weaknesses in the wording of the 1901 Act and involved no significant change in mining administration.[88]

The absence of significant new initiatives in government mining administration in these decades should not be seen as evidence of a decline in Bangkok's interest in the peninsular tin industry. Rather, it indicates a recognition on the part of the central government that further development of its mining administration would involve a far greater measure of direct intervention in the industry than had occurred thus far. In the two decades following the establishment of the Department of Mines in 1892, the government had created the basic administrative and legal structures that enabled it to regulate the activities of tin mining concerns in the peninsula. That regulation was essential if the central government was to secure social and political stability in that distant region of the kingdom in a period when it was experiencing substantial, but potentially disruptive, economic change. But at the same time, those administrative and legal structures gave Bangkok little or no influence in determining the principal features of the peninsular tin industry in the late nineteenth and early twentieth centuries—its size and rate of growth; its ethnic structure; the pace of technological change within it; and its high and increasing dependence on British interests in Malaya for the smelting of its output. The important influences here included the relative attractiveness of other tin-producers for investment by Western capital; the changing technological demands of tin mining in Siam; the level of world demand for tin; and the economic advantages secured through large-scale tin smelting. Thus, for example, although Crown Prince Vajiravudh and Čhaophrayā Wongsānupraphat may have wished to encourage a more rapid growth of the peninsular tin industry, in practice this was essentially beyond the influence of government in this period. And indeed, although tin production in Siam continued its modest growth through the 1910s, it fell back considerably at the beginning of the 1920s, to recover only at the very end of that decade.[89] Neither could the government in this period, if it had wished, significantly reduce the relative importance of Western capital in the industry, accelerate the pace of technological change, or remove the dependence on the Penang smelters. To have done so would have required a very major investment of capital, and of administrative and technical expertise on the part of the government; and almost certainly would have involved the kingdom in denying Western capital rights which had been secured by treaty in the mid-nineteenth century. In brief, given the openness of the peninsula to Western investment and trade, and the very limited resources at the command of the Siamese administration in the opening decades of the twentieth century, Bangkok could only regulate the tin industry; it could not shape it.[90]

IV

As was noted in the opening of this chapter, the extraction of teak for export from the forests of northern Siam in the late nineteenth and early twentieth centuries confronted the Bangkok administration with problems sharply different from those which it faced with respect to the peninsular tin industry in the same period. To a large degree that difference derived from the greater involvement of Western capital in the exploitation of Siam's teak than in the extraction of its tin.

Western commercial interest in Siam's teak stands dates from the beginning of the 1860s when the Borneo Company, which was to emerge as one of the major timber companies in the region, placed agents in the north of the kingdom.[91] However, at that time conditions in the teak districts were not conducive to a large-scale working of the forests. The principal difficulty was that there were no secure procedures to govern the granting of forest leases, and no regulation at all of the actual working of the forests.[92] Consequently, there were continuous disputes over forest rights between leaseholders and the forest-owners, the local princes. During the 1860s and early 1870s these disputes acquired increasingly prominent political implications, for a growing number of leaseholders were Burmese foresters from British-ruled Lower Burma. Thus in 1871 a number of Burmese foresters, claiming extraterritorial rights, brought their grievances against the northern princes to the British Consular Court in Bangkok.[93] In the course of the proceedings, serious irregularities in the granting of forest concessions were revealed. Three years later, the Siamese Government moved to reduce the possibility of conflict with Britain arising from the working of the northern forests by Burmese leaseholders. Under the Anglo-Siamese Treaty of 1874, Bangkok became responsible for the adjudication of disputes between Asian subjects of Britain and citizens of Chiangmai, Lampāng, and Lamphūn.[94] In the same year, Bangkok appointed a commissioner to Chiangmai to enforce the treaty and to secure for central government part of the forest revenues raised in the north, and introduced legislation which required forest agreements between the northern princes and foreign subjects to be ratified by the Bangkok authorities.[95] A further Anglo-Siamese Treaty in 1883 transferred jurisdiction over disputed leases to the newly-established British Consulate at Chiangmai and, for the first time, made provision for Europeans themselves to undertake the actual working of the forests (although British subjects seeking to work the forests in the north were first required to secure permission from Bangkok).[96] The following year, 1884, the Bangkok administration, acting through its commissioner in Chiangmai, began to dictate the form of forest lease to be used in the north.[97]

The major expansion in teak extraction from northern Siam which was to secure for the kingdom a dominant position in this trade by the early twentieth century, took place from the mid-1880s. It was initially stimulated by the over-exploitation of the Burmese teak forests and then the imposition by the colonial administration of strict regulations over felling in Upper Burma immediately following Britain's annexation of the territory in 1886, both of which drove labour and capital into the teak districts of Siam.[98] In the early years of the expansion, the actual forest operations remained in the hands of Burmese and Shans, the European timber companies with offices in Chiangmai and Moulmein advancing them the major loans necessary for this work.[99] However, towards the end of the 1880s, those companies began to take on forest operations themselves in order to protect their substantial investment in the industry and to secure their supply of high quality teak. Thus in 1888 the Borneo Company became the first European firm to obtain a forest lease. By the end of the following decade, it had been joined in working the teak forests by the Bombay Burmah Trading Corporation, the Anglo-Siam Corporation, Louis T. Leonowens, the Danish East Asiatic Company, and the French East Asiatic Company (the last floating its teak down the Mekong to Saigon).[100] Even so, as late as 1895 most forest leases, although presumably not necessarily the most extensive ones, were still held by Burmese foresters.[101]

The major expansion in teak extraction (the annual average volume of teak exports through Bangkok rose from 5 600 cubic metres in 1873–6 to 37 000 in 1890–4),[102] and the emerging dominance of Western capital in the industry, considerably disturbed the Bangkok administration—for it was clear that the measures taken in the 1870s and early 1880s, notably the appointment of a commissioner to Chiangmai and the treaty agreements with Britain, were now insufficient if the central government was to maintain control of developments in the northern teak forests. Bangkok's initial response was to establish a substantial commercial interest of its own in teak extraction. This was achieved in partnership with an American doctor, Marion A. Cheek, who for many years had been a member of the Presbyterian Mission in Chiangmai.[103] Encouraged by the rapid expansion of the teak industry from the mid-1880s, Cheek had resigned from the Mission and devoted all his capital and time to building up a teak business. However, he was soon in financial difficulty. He had borrowed heavily, but had used his profits primarily to purchase more elephants, forestry equipment, and leases, so leaving himself with insufficient funds in hand to work all his forests and in this way secure the income to pay off his debts. The fact that it took on average five years for an investment in the industry to show a return, this being the time that would elapse between the felling of a teak tree in the north and its arrival in Bangkok and sale,[104] further weakened his financial position. By the end of 1888, Cheek owed the Borneo Company some 250,000 baht and had just 25,000 baht in hand. He then approached the Siamese Government for assistance. He was well received, for the administration saw in support for Cheek a means of counter-balancing the increasingly powerful position of the Borneo Company and the Bombay Burmah Trading Corporation in the teak industry and of encouraging American diplomatic support against a possible British forward movement into northern Siam. In April 1889, an agreement was concluded between Cheek and Prince Narāthip (the head of the embryo Ministry of Finance), representing the Siamese Government, under which the former was advanced 600,000 baht 'to be used in the working of teak forests and the purchasing of teak wood'.[105] The loan attracted interest at the rate of 7.5 per cent per annum, and was made on the security of all the teak wood and elephants owned by Cheek during the period of the agreement. The working of the forests and the sale of the felled teak were left solely in Cheek's hands, but all his forest leases were to become the property of Prince Narāthip, again for the period of the agreement. Of the net profit arising from Cheek's business, two-thirds would be retained by Cheek, the remaining one-third would go to Prince Narāthip; the latter would not be liable for losses incurred by the business. In January 1890 the Siamese Government advanced Cheek a further 200,000 baht on identical terms.

By the following year, Cheek was in serious conflict with his Siamese creditors. Because of poor rains in 1890, relatively little teak could be floated down to Bangkok for sale during that season, and consequently in March 1891, Cheek was unable to make an interest payment on his loan. The following floating season was even worse, so that in March 1892 he again defaulted. In August of that year, Prince Devawongse declared that as Cheek was in breach of his agreement, the Siamese Government, acting through an official receiver, would assume ownership of all his teak rafts coming down to Bangkok; and in July 1893, finding that Cheek was continuing his operations in the north despite this action against him, the government posted a public notice throughout the forest districts around Chiangmai which, in effect, placed an injunction on the use of his elephants and equipment. These events provoked lengthy argument between the Siamese authorities and the American Legation in Bangkok, the details of which are of little concern in this context.[106] It need only be noted that eventually, in early 1897, the parties agreed to place the dispute before an arbitrator, Sir Nicholas Hannen, who was Chief Justice of the British Consular Court for China and Japan, and British Consul-General at Shanghai. Hannen's judgment, announced in March 1898, was in favour of Cheek, or rather his estate, Cheek having died in 1895. It was decided that the Siamese Government's seizure of Cheek's property from August 1892 was a violation of the United States–Siam Treaty of 1856; and, moreover, that Cheek had not been in default with respect to the payment of interest since the agreement with Prince Narāthip had simply stated the annual rate of interest to be applied to the loan, and in so doing had not required (even by implication) that that interest be paid each year. Hannen's arbitration involved the Siamese Government in paying an indemnity of 706,721 baht to Cheek's estate in order to restore it to its value prior to August 1892.

The rapid collapse of the government's partnership with Cheek, and thus of its attempt to create a counter-balance to the increasingly powerful British timber companies in the north, undoubtedly encouraged the administration in the mid-1890s in its decision to impose far stricter control over the working of the northern teak forests. It was not the only consideration. Also important here was the increasingly arbitrary and extortionate behaviour of the northern princes, for as the owners of the forests they were able to exploit the heightened competition between the Western timber companies to extract a widening range of charges.[107] Inevitably, complaints by the companies against the obstructive, demanding manner of the princes were forcefully made to the authorities in Bangkok. A further important influence was the realization on the part of the central government that under the existing lax administration of the forests, the teak resources of the kingdom would soon be exhausted. Thus when Bangkok introduced a new form of forest lease through the commissioner in Chiangmai in 1893, in place of that in use from 1884, it included the requirement that the lessee plant saplings to replace trees felled.[108] A final consideration was the failure of the central government in this period to receive its agreed share (or indeed, in some cases any part) of forest revenues from the northern princes.[109] Here was a particularly strong reason for a more interventionist forestry administration from Bangkok.

To prepare the way for this major shift in administration, in 1895 the government secured the services of a European forestry official, H. A. Slade. This was a most astute appointment. As Slade had been trained at the French Forest School at Nancy and then appointed to the Imperial Forest Service in Burma,[110] French and British sensibilities were assuaged. More importantly, there were clear advantages for the Siamese authorities in their employment of an experienced British official to impose strict regulation on an industry dominated by British capital. Slade arrived in Bangkok in January 1896, and after completing an extensive tour of the teak districts submitted a report to the Ministry of the Interior in August of that year.[111] Among his principal proposals was that ownership of the northern teak forests be transferred from the local princes to the central government; with this the princes would lose their claim on the forest revenues but would instead receive a regular payment from the government. Slade further proposed the creation of a specialist Forestry Department; that each year two or three Siamese officials be sent abroad for specialist instruction in forestry; that legislation be introduced to regulate strictly all forestry operations; and that a more restrictive form of forest lease also be introduced.

The Siamese Government moved quickly to implement Slade's proposals. Within a few weeks of the submission of his report, a Forestry Department had been created in the Ministry of the Interior.[112] The Department was located in Chiangmai, and Slade himself was appointed its first Director.[113] The new Department immediately embarked on a major reorganization of forest administration. In 1897, the central government assumed ownership of the northern teak forests, the local princes having been encouraged to relinquish their rights by the provision of favourable financial settlements.[114] In the same year, a new form of forest lease was introduced which imposed considerably more restrictive conditions for the felling of teak.[115] A further new form of lease was brought into force three years later, in 1900.[116] This provided for the closing of one half of the teak districts for a period of 15 years, prohibited lessees from undertaking girdling themselves, and raised sharply the royalty to be paid on logs extracted. During this early period, the Forestry Department was also involved in preparing legislation designed to curb various malpractices long found in the working of the teak forests.[117] Important here was legislation brought into force at the beginning of 1897 which, seeking to reduce the incidence of timber theft, contained provision for forestry officials to impound all timber where the owner's distinguishing hammer-mark had been defaced, and imposed heavy penalties on any person found altering those marks.[118] It was also during these early years, under Slade, that the Forestry Department sought to create the administrative structure and procedures through which it could effectively regulate teak extraction throughout the north. Thus a survey of all open forests was undertaken by the Department's officers, and periodical inspection of the timber companies' operations was instituted to ensure that lease conditions were being strictly observed.[119] In 1898, the government's teak duty station was moved northwards from Chaināt to the confluence of the Ping, Wang, Yom, and Nān at Pāknamphō, for it had been found that many of the teak logs being floated out had escaped the payment of duty by being removed from the river and sold before they reached Chaināt.[120] In the same year, a Siamese official was posted to Moulmein district in Burma to collect duty on the large volume of teak which was felled in northern Siam but which was being floated down the Salween rather than the Čhaophrayā.[121] Finally, in 1901 the Forestry Department sent its first group of Siamese officials abroad for specialist training, to the Imperial Forest College at Dehra Dun in India.[122] On their return they moved into senior positions in the Department.

Slade left the Siamese service in 1901 and returned to Burma.[123] The changes in forest administration which had been introduced under his direction from 1896 represented a considerable achievement, particularly when compared with the performance of the Department of Mines in this period. The greater incisiveness of the Forestry Department derived primarily from the fact that the need for the Bangkok administration to establish control over the extraction of teak was considerably more compelling than the need for it to impose order on the exploitation of tin. The argument here relates primarily to the contrasting scale of Western investment in the two industries. It is not simply the fact that from the 1890s teak extraction was dominated by a small number of European timber companies, among which British firms (notably the Borneo Company and the Bombay Burmah Trading Corporation) were the most powerful, while tin mining attracted no significant Western capital until 1907 and remained largely controlled by Chinese interests until the 1920s. More important is the observation that while Western timber companies in the region focused their activities on the teak districts of northern Siam from the final decades of the nineteenth century, in the same period Western mining capital showed relatively little interest in the kingdom but was overwhelmingly attracted into the British territory to the south. In brief, Siam was far more important to Western timber firms than to Western mining interests. It must be assumed that it was considerations such as these which prompted the government to locate the Forestry Department not in the capital but in the heart of the teak region itself.[124] But if the need for the assertion of central government authority was more urgent in the case of teak, in many respects it was also easier to achieve. The authorities in Bangkok may have been disturbed to see logging in the north dominated almost exclusively by a handful of powerful Western companies, but that concentration of interests did at least facilitate the negotiation of an effective regulation of teak extraction. More importantly the companies themselves accepted the need for government-imposed restriction. It is true that in the mid-1890s they 'indiscriminately girdled or felled every tree they could lay hands on' in anticipation of an imminent curtailment of felling rights;[125] and at the end of that decade the Bombay Burmah Trading Corporation negotiated fiercely with Slade over the renewal of its forest leases.[126] Nevertheless, even though the Western timber companies occasionally sought short-term advantage in these ways, they did not challenge the administration's determination from the close of the nineteenth century to restrict felling to the rate at which the teak stands could be renewed, for they well realized that the long-term prosperity of the industry (if not its actual existence), and hence their own future, depended on effective measures of forest conservation. In this important respect there was an essential unity of interest between the Siamese Government and the Western forest companies.

Finally, the incisiveness of the government's forest administration in this period undoubtedly also reflected in part the strong temperament and commitment of the European officials appointed to the Forestry Department. Slade was clearly an imposing influence, a man 'of exceptional ability and force of character' according to one near-contemporary,[127] who dealt very firmly with his compatriots in the timber companies during the negotiation of the new leases from 1897.[128] It is therefore significant that while the Department of Mines had a Siamese director from 1907, as a rift had opened between the senior European officials in the Department and Prince Damrong, the Forestry Department retained a British director at least until the mid-1920s.[129] It is not easy to explain the greater commitment of the European forestry officials. Of course the essential direction which forest administration would take in this period—to restrict the operations of the Western timber companies to allow for forest renewal—was beyond challenge, even from the companies themselves. Yet it would be surprising if part of the explanation did not lie simply with the nature of forest work and with the character and outlook of the men it attracted.

The measures introduced under Slade's direction in the closing years of the nineteenth century established the essential features of forest administration in Siam to the end of the 1930s. When most of the teak leases negotiated by Slade expired in 1909, the Forestry Department carried out a major consolidation of lease areas (thus reducing the number of leases from 105 to 40), extended the period of the lease from 6 years to 15 years, and raised the royalty on teak timber by 20 per cent.[130] A further revision of leases in 1925 reduced their number to 33, raised the minimum girth at which a teak tree could be felled, and removed the actual selection and girdling of the teak trees from the lessees to the officials of the Forestry Department. The imposition of increasingly tight restriction on the operations of the Western timber companies from the end of the nineteenth century led, as of course was intended, to a notable contraction in the rate at which teak was extracted. Teak exports reached their highest level in the mid-1900s, reflecting the very high rate of felling just prior to the negotiation of the 1897 leases. Average annual export in 1905–7 was 134 877 cubic metres.[131] In 1925–7 it was 81 373 cubic metres.

A more important effect of the increasingly restrictive working conditions imposed by the Forestry Department in this period was, according to one contemporary, 'to throw the teak industry more and more into the hands of the big firms, to the exclusion of the small foresters'.[132] The principal argument here is that the institution of large lease-areas (notably so with the consolidation of leases in 1909), within which only selected trees of a minimum girth could be felled, raised the labour requirements of forest work far above the resources of the individual Burmese or Shan forester.[133] It was also argued that a number of the detailed regulations introduced from the end of the nineteenth century, for example that which imposed a fine for each teak tree so damaged in felling as to be not worth extracting, bore more heavily on the individual Asian forester than on the highly-capitalized Western company;[134] and that the steep increases in fees and royalty payments imposed in this period pared away the slim profit margins of the Burmese and Shan operators.[135] However, the government's forestry measures should be seen as simply encouraging the emergence of a Western oligopoly in the Siamese teak industry, but certainly not its primary cause. The central consideration here was the fact that the efficient large-scale working of the teak stands of northern Siam demanded an investment of fixed and working capital on a scale that only the major Western companies could provide. This was particularly the case as the teak stands near the main streams became exhausted and logging had to move deeper into the forests, for this more remote working required heavy investment in elephants, labour, and forestry equipment.[136]

The more interesting question concerns not the government's role in strengthening the Western domination of the teak industry but, on the contrary, its failure to challenge it. Although the

THE ENCLAVES: TIN AND TEAK 119

extraction of teak by the Forestry Department itself was undertaken (initially in the Māe Hāet district of Phrāe) from 1912,[137] by 1930 the Department was reponsible for working just 1 per cent of the teak forests, the European firms accounting for 85 per cent and local lessees for the remaining 14 per cent.[138] It is possible that the Siamese administration was discouraged from establishing a major direct interest in the extraction of teak by the failure of its partnership with Cheek in the early 1890s. The point must also be made that with the introduction of long leases in 1909 the rights of the Western companies to work the northern forests were in effect secured to the end of the 1930s. Yet, it is clear that the principal reason why the Siamese administration made no attempt to break the Western domination of the industry in the opening decades of the twentieth century was simply because in this period it could command neither the capital resources nor the trained manpower to make government exploitation of the teak forests possible.[139] In this crucial respect, government administration of the teak industry in the late nineteenth and early twentieth centuries ran closely parallel to that of the tin industry in the same period. For, in essence, government sought simply to regulate the northern industry: it did not seek to restructure it.

1. It is not possible to provide precise figures for the relative importance of these three exports, partly because there were very substantial annual variations in export values for each commodity but particularly because (for reasons to be noted below) official data on tin and teak exports in this period are incomplete. For an outline of the data and a brief discussion of its imperfections, see, James C. Ingram, *Economic Change in Thailand 1850-1970*, Stanford, 1971, pp. 93-7.
2. These observations draw on unpublished figures for tin output in Siam (1845-50 to 1914) prepared by M. E. Falkus, primarily from the data on tin imports from the kingdom into Penang. It should be noted here that Siam's official customs figures in the period before 1920 apply only to the port of Bangkok; tin exports from the southern provinces did not, of course, pass through that port.
3. Unpublished figures prepared by M. E. Falkus.
4. Wong Lin Ken, *The Malayan Tin Industry to 1914: With Special Reference to the States of Perak, Selangor, Negri Sembilan and Pahang*, Tucson, 1965, p. 246.
5. H. Warington Smyth, *Five Years in Siam: From 1891 to 1896*, London, 1898, vol. 1, pp. 318, 320.
6. Great Britain, Foreign Office, *Diplomatic and Consular Reports: Siam* (henceforth cited as *DCR*), no. 3999, 1908 [for 1907], p. 4.

7. See for example, Phrayā Surasakmontrī to Prince Sommot, 19 October 1896, NA r5 KS 6.5/8.
8. *DCR*, no. 3788, 1907 [for 1906], p. 7.
9. Ibid.; J. W. Cushman, 'The Khaw Group: Chinese Business in Early Twentieth-century Penang', *Journal of Southeast Asian Studies*, vol. 17, no. 1 (March 1986), pp. 67, 71–2.
10. *DCR*, no. 4622, 1911 [for 1909–10], p. 4.
11. G. William Skinner, *Chinese Society in Thailand: An Analytical History*, Ithaca, 1957, p. 110.
12. Phrayā Thipkōsā to Prince Narāthip, 16 March 1893, NA r5 KS 6.1/6; Cushman, op. cit., p. 71.
13. Skinner, op. cit., p. 110.
14. 'Thīprachum kamakānphisēt truat phrarātchabanyatrāe naikrasuangkasēt' ('Meeting of the special committee to examine the mining legislation, held in the Ministry of Agriculture'), 1 April 1896, NA r5 KS 6.2/1.
15. Phrayā Thipkōsā to Prince Narāthip, 16 March 1893, NA r5 KS 6.1/6.
16. Chulalongkorn to Čhaophrayā Thēwēt, 29 November 1896, NA r5 KS 6.2/1. The King was probably referring here to a Labour Code, introduced in the Malay States in 1895, which 'for the first time, defined exactly the duties and powers of the magistrates, police, and the Protector of Chinese, and the rights and obligations of employers and labourers'. (See Wong, op. cit., p. 73.) However, it should be added that, again according to Wong (p. 74), 'the labour laws were honoured more in the breach than in the observance'.
17. Walter de Müller to Phra Prachāchīp, 7 January 1895, NA r5 KS 6.5/8.
18. Smyth, op. cit., vol. 1, p. 318.
19. Walter de Müller to Phrayā Surasakmontrī, 28 November 1894, NA r5 KS 6.5/8.
20. Walter de Müller to Phra Prachāchīp, 7 January 1895, NA r5 KS 6.5/8.
21. Wong, op. cit., p. 255.
22. Prince Damrong to Chulalongkorn, 15 January 1902, NA r5 KS 6.1/32.
23. Smyth, op. cit., vol. 1, p. 319.
24. Chulalongkorn to Čhaophrayā Thēwēt, 29 November 1896, NA r5 KS 6.2/1; Prince Damrong, 'Rāingān truat huamu'ang phaktāi r.s. 115' ('Report on an inspection tour of the southern provinces, 1896'), NA r5 M 2.14/74, translated in Chatthip Nartsupha and Suthy Prasartset (eds.), *The Political Economy of Siam 1851–1910*, Bangkok, 1981, pp. 418–25.
25. H. Warington Smyth to Phrayā Surasakmontrī, 15 September 1896, NA r5 KS 6.2/2.
26. Čhaophrayā Wongsānupraphat, *Prawat krasuangkasētrāthikān (History of the Ministry of Agriculture)*, Bangkok, 1941, p. 162.
27. 'Thīprachum kamakānphisēt', 1 April 1896, NA r5 KS 6.2/1.
28. Smyth, op. cit., vol. 1, p. 320.
29. Chulalongkorn to Čhaophrayā Thēwēt, 29 November 1896, NA r5 KS 6.2/1.
30. Čhaophrayā Wongsānupraphat, op. cit., p. 163.
31. Ibid., p. 165.
32. W. A. Graham, *Siam*, London, 1924, vol. 1, p. 360.
33. Tej Bunnag, *The Provincial Administration of Siam 1892–1915: The Ministry of the Interior under Prince Damrong Rajanubhab*, Kuala Lumpur, 1977, pp. 268, 275.

34. Ministry of Agriculture, *Prawat krasuangkasēt* (*History of the Ministry of Agriculture*), Bangkok, 1957, p. 85; Tej, *The Provincial Administration of Siam 1892–1915*, p. 97.
35. Tej, *The Provincial Administration of Siam 1892–1915*, pp. 223–4.
36. Čhaophrayā Wongsānupraphat, op. cit., p. 165; Prince Devawongse to the foreign representatives in Bangkok, March 1897, NA r5 KS 6.2/1.
37. 'Thīprachum kamakānphisēt', 1 April 1896, NA r5 KS 6.2/1.
38. H. Warington Smyth to Phrayā Surasakmontrī, 15 September 1896 and 18 September 1896, NA r5 KS 6.2/2.
39. H. Warington Smyth to Phrayā Surasakmontrī, 15 September 1896, NA r5 KS 6.2/2.
40. Ibid.; Prince Mahit to Chulalongkorn, 20 October 1896, NA r5 KS 6.2/2.
41. Chulalongkorn to Čhaophrayā Thēwēt, 29 November 1896, NA r5 KS 6.2/1.
42. Documents in NA r5 KS 6.2/1.
43. Čhaophrayā Wongsānupraphat, op. cit., pp. 171–2.
44. Unpublished figures prepared by M. E. Falkus.
45. Royal Department of Mines, *Notes on Mining in Siam*, Bangkok, 1921, pp. 2–3. This was the company which in 1906 undertook to dredge an entrance to the harbour and a dock at Phūket. A brief history of the early years of the Tongkah Harbour Tin Dredging Company by its founder, Captain E. T. Miles, originally published in 1919, is reproduced in, Chatthip and Suthy, op. cit., pp. 251–64.
46. Ministry of Commerce and Communications, *Siam: Nature and Industry*, Bangkok, 1930, p. 114.
47. H. Warington Smyth to Phrayā Surasakmontrī, 15 September 1896, NA r5 KS 6.2/2.
48. Royal Department of Mines, op. cit., p. 3; Ministry of Commerce and Communications, op. cit., p. 114.
49. It has been argued that the Chinese entrepreneurs in the peninsula 'lacked the capital to purchase dredging equipment' (Skinner, op. cit., p. 215). The identical argument has been advanced with respect to the Malay States: 'bucket dredges were beyond the financial means of the traditional Chinese mining organizations' (Wong, op. cit., p. 210). It is true that the purchase and erection of a bucket dredge demanded a major capital investment. (The cost of floating the Tongkah Harbour Tin Dredging Company in 1906, which had one dredge, was £17,500: see Chatthip and Suthy, op. cit., pp. 257–8.) But the Chinese mining capitalists here were men of considerable wealth, varied business interests, and extensive commercial connections with other Chinese entrepreneurial groups in the region, such that it is difficult to accept that they could not command the resources to make this investment. This is an important issue which clearly requires further research.
50. Cushman, op. cit., p. 63.
51. Smyth, op. cit., vol. 1, p. 328.
52. Wong, op. cit., pp. 163–4, 229.
53. Cushman, op. cit., pp. 72–5.
54. Documents in NA r5 KS 6.5/3.
55. The process by which the Straits Trading Company also eclipsed its small-scale Chinese rivals in the Malay States in this period is considered in Wong, op. cit., pp. 165–7, 227–8.
56. *DCR*, no. 4622, 1911 [for 1909–10].
57. Royal Department of Mines, op. cit., p. 1.

58. Skinner, op. cit., p. 215.
59. In the period 1910–14 the average annual output of tin from Siam was just one-fifth of average annual production in Perak alone. (Unpublished figures prepared by M. E. Falkus; Wong, op. cit., p. 249.)
60. Tej, *The Provincial Administration of Siam 1892–1915*, pp. 97, 226.
61. Prince Damrong to Chulalongkorn, 12 December 1901, NA r5 KS 6.1/32.
62. Walter de Müller to Phrayā Surasakmontrī, 28 November 1894, NA r5 KS 6.5/8.
63. Prince Damrong to Chulalongkorn, 28 March 1895, NA r5 KS 6.5/8.
64. Smyth, op. cit., vol. 1, pp. 328–30.
65. Documents in NA r5 KS 6.5/35.
66. Chulalongkorn to Prince Damrong, 1 March 1906, NA r5 KS 6.5/35. The previous year the superintendent commissioner in Phūket had been authorized by Bangkok to promulgate emergency regulations in order to deal with violent labour disputes in the local tin-mines. (Tej, *The Provincial Administration of Siam 1892–1915*, p. 173.)
67. Prince Mahit to Chulalongkorn, 23 December 1898, NA r5 KS 6.2/1.
68. Chulalongkorn to Prince Mahit, 27 December 1898, NA r5 KS 6.2/1.
69. Prince Damrong, 'Ru'ang kānthamrāe' ('Memorandum on Mining'): enclosed in, Prince Damrong to Chulalongkorn, 25 August 1909, NA r6 KS 6/2.
70. This phrase had been used in a comparable context by H. Warington Smyth (Smyth, op. cit., vol. 1, p. 299).
71. The following draws on, Thamsook Numnonda, 'The Anglo-Siamese Secret Convention of 1897', *Journal of the Siam Society*, vol. 53, pt. 1 (January 1965), pp. 45–60.
72. David F. Holm, 'The Role of the State Railways in Thai History, 1892–1932', Ph.D. diss., Yale University, 1977, pp. 145–6.
73. Prince Damrong, 'Ru'ang kānthamrāe' ('Memorandum on Mining'), NA r6 KS 6/2.
74. Meeting of the Council of Ministers, 26 August 1909, NA r6 KS 6/2.
75. Prince Damrong, 'Ru'ang kānthamrāe' ('Memorandum on Mining'), NA r6 KS 6/2.
76. Ibid.
77. Ministry of Agriculture, op. cit., pp. 87–8.
78. Prince Damrong, 'Ru'ang kānthamrāe' ('Memorandum on Mining'), NA r6 KS 6/2.
79. Prince Damrong to Chulalongkorn, 16 October 1907, NA r5 KS 6.1/32.
80. Ministry of Agriculture, op. cit., p. 88.
81. Tej, *The Provincial Administration of Siam 1892–1915*, pp. 226, 244.
82. Čhaophrayā Wongsānupraphat, 'Nai kānkhuthā nammandin lae thānsilā' ('Memorandum on prospecting for coal tar and coal'), October 1911, NA r6 KS 6/2.
83. Documents in NA r6 KS 6/2.
84. Koh Tiew Lim to Vajiravudh, 20 April 1918, NA r6 KS 6/9. This document appears in an English translation in, Chatthip Nartsupha, Suthy Prasartset and Montri Chenvidyakarn (eds.), *The Political Economy of Siam 1910–1932*, Bangkok, 1981, pp. 80–4.
85. Phrayā Thipkōsā to Prince Devawongse, 29 January 1893, NA r5 KS 6.5/6.
86. Documents in NA r6 KS 6/9.
87. It was not until the mid-1960s that a modern tin smelter was constructed in

Phūket. (Wolf Donner, *The Five Faces of Thailand: An Economic Geography*, London, 1978, p. 527.)

88. Ministry of Agriculture, op. cit., pp. 117–18.
89. Constance M. Wilson, *Thailand: A Handbook of Historical Statistics*, Boston, 1983, p. 153.
90. The point should be made here that it was not until the 1970s that the Thai Government took effective action with respect to one of the most prominent features of the peninsular tin industry throughout the modern period, that is, its domination by foreign capital. In the mid-1970s the four principal tin producers in the kingdom comprised an international company based in Holland and three Malaysian companies; and until 1978 most tin concessions in the peninsula were worked by foreign companies with only minority Thai interests. However, in that last year 'the Thai government passed new legislation limiting mining concessions to majority Thai-owned companies incorporated in Thailand'. (Wilson, op. cit., p. 148.)
91. Virginia Thompson, *Thailand: The New Siam*, New York, 1941, p. 833.
92. Ministry of Agriculture, op. cit., p. 149.
93. Tej, *The Provincial Administration of Siam 1892–1915*, p. 68.
94. Ibid., pp. 69–70.
95. Ministry of Agriculture, op. cit., p. 149.
96. David F. Holm, 'A History of the Teak Industry in Thailand', unpublished paper, Yale University, May 1969, p. 7; Ingram, op. cit., p. 106; Ministry of Agriculture, op. cit., pp. 149–50.
97. Holm, 'A History of the Teak Industry in Thailand', p. 12.
98. Ibid., p. 7; Ingram, op. cit., p. 106.
99. Holm, 'A History of the Teak Industry in Thailand', pp. 6–8, 18.
100. Reginald le May, *An Asian Arcady: The Land and Peoples of Northern Siam*, Cambridge, 1926, p. 58.
101. Ingram, op. cit., p. 106.
102. Ibid., p. 96.
103. The following draws on James V. Martin, 'A History of the Diplomatic Relations between Siam and the United States of America, 1833–1929', Ph.D. diss., Fletcher School of Law and Administration, 1947, chapter 6.
104. Ministry of Commerce and Communications, op. cit., p. 132.
105. Martin, op. cit., p. 209.
106. A lengthy account, essentially sympathetic to Cheek, is provided in ibid., chapter 6.
107. Ministry of Agriculture, op. cit., p. 150.
108. Smyth, op. cit., vol. 2, p. 280.
109. Tej, *The Provincial Administration of Siam 1892–1915*, pp. 70, 143.
110. Le May, *An Asian Arcady*, pp. 61–2.
111. Ministry of Agriculture, op. cit., pp. 150–2.
112. Ibid., pp. 152–3.
113. Tej, *The Provincial Administration of Siam 1892–1915*, p. 96.
114. Ministry of Agriculture, op. cit., p. 153.
115. Ibid., pp. 153–4; Ministry of Commerce and Communications, op. cit., pp. 126–7.
116. Le May, *An Asian Arcady*, p. 61.
117. Ministry of Commerce and Communications, op. cit., p. 127.
118. Smyth, op. cit., vol. 1, p. 71.

119. Le May, *An Asian Arcady*, p. 61.
120. Ministry of Agriculture, op. cit., p. 154.
121. Ibid. In the early twentieth century almost one-fifth of all teak logs brought out of northern Siam were floated down the Salween. (Ministry of Commerce and Communications, op. cit., p. 130.)
122. Ministry of Agriculture, op. cit., p. 155.
123. Le May, *An Asian Arcady*, p. 62.
124. The Department was transferred from Chiangmai to Bangkok in 1910 (Tej, *The Provincial Administration of Siam 1892–1915*, p. 223), that is, after effective government regulation of teak extraction had been established.
125. Holm, 'A History of the Teak Industry in Thailand', p. 13.
126. A. C. Pointon, *The Bombay Burmah Trading Corporation Limited 1863–1963*, London, 1964, pp. 37–8.
127. Le May, *An Asian Arcady*, p. 61.
128. Pointon, op. cit., pp. 37–8.
129. W. F. Lloyd, seconded from Burma, was Director of the Forestry Department from 1905 to 1924. (Ministry of Agriculture, op. cit., p. 155.)
130. Ministry of Commerce and Communications, op. cit., pp. 127–9.
131. Wilson, op. cit., p. 213.
132. Smyth, op. cit., vol. 2, p. 283.
133. Ingram, op. cit., pp. 106–7; Holm, 'A History of the Teak Industry in Thailand', p. 23.
134. Smyth, op. cit., vol. 2, p. 283.
135. Holm, 'A History of the Teak Industry in Thailand', pp. 22–3.
136. Ibid., p. 23.
137. Ministry of Agriculture, op. cit., p. 156.
138. Ministry of Commerce and Communications, op. cit., p. 129.
139. Thompson, op. cit., p. 473. This is an appropriate point to add that it was not until 1940 that the Siamese Government undertook to work a major part of the teak forests (Holm, 'A History of the Teak Industry in Thailand', p. 57, note 74); and not until the late 1950s that the final interest of the Western timber companies in working the northern forests was removed (ibid., p. 46).

4
The Élite and the Early Development of Indigenous Banking

IN the final two chapters of this study, the focus moves from the major export sectors of the Siamese economy principally to a consideration of the emergence of new commercial and manufacturing interests in the kingdom from the early twentieth century, and from the Bangkok political élite as administrators to the élite primarily as entrepreneurs. The present chapter is concerned with the establishment and early years of the Siam Commercial Bank, the first indigenous modern bank in the kingdom and one of the major banking institutions in present-day Thailand.

I

Within a few years of the opening of Siam to unrestricted foreign trade in the 1850s, the two principal British banks in the East, the Hongkong and Shanghai Banking Corporation and the Chartered Bank of India, Australia and China, had appointed agents in Bangkok.[1] In 1888, the Hongkong and Shanghai became the first Western bank to establish a branch office in Siam.[2] It was followed in 1894 by the Chartered Bank and in 1897 by the Banque de L'Indochine. These branch offices assumed three principal functions in this early period. First, they acted as deposit banks. Particularly important in this respect was the Hongkong and Shanghai Bank. As early as 1890, that bank had attracted some 355 current accounts; its clients included the principal foreign companies in Bangkok, members of the Siamese élite including the King himself, as well as the Customs Department and the Treasury.[3] Second, the three branch offices acted as banks of issue. Here again the Hongkong bank was particularly important, being the first to issue notes (in 1889) and accounting for the major portion of the total banknote issue at the close of the century.[4] However, even that bank's notes did not gain a wide circulation, and when the Siamese administration introduced its own paper currency in 1902, the foreign issues were rapidly withdrawn. But the principal function of the European banks in Bangkok in this period was of course the conduct of exchange operations, the major part of which consisted of the purchase from local Chinese rice millers of dollar bills of exchange drawn against shipments of rice to Hong Kong and Singapore.[5] In effect, from the end of the nineteenth century the export of the rice crop rested upon the exchange facilities provided by the European banks. Here lay a potentially major source of power for them—and in late 1902 they saw reason to exercise it.

On 25 November 1902, the Siamese administration closed the mints to the free coinage of silver and placed the baht on a gold-exchange standard, in the process revaluing the currency against sterling by approximately 23 per cent.[6] These measures provoked a storm of protest from the three European banks in Bangkok. Each held a substantial part of its deposits in Mexican dollars in Hong Kong and Singapore and consequently, with the revaluation of the baht, stood to lose heavily when they came to reimport those funds into the kingdom and exchange them again for the local currency. The managers of the banks saw the Financial Adviser, Rivett-Carnac, on 27 November to ask whether the government would compensate them for these losses. Rivett-Carnac refused to consider their claims. The response of the Hongkong and Shanghai Bank and of the Chartered Bank was immediate and emphatic—they threatened to refuse to purchase baht at the new rate announced by the authorities, and indeed it would appear that at one point they carried out that threat and refused all exchange business. In addition, according to the Minister of Finance, Prince Mahit, a meeting was called between the British banks and the principal rice millers in the port at which the former sought to have the export of the rice crop held back. Within two weeks the Siamese Government gave way. Under an agreement with the banks, eventually concluded on 19 December 1902, the administration revalued the baht by a considerably smaller degree than had been decided at the end of November, and gave an assurance that future revaluations of the currency would be gradual in order to minimize disturbance to trade. In addition, the banks were allowed to import Mexican dollars and exchange them for baht at the rate prevailing before the mints were closed, up to the value of government deposits with them on 26 November 1902. In brief, by threatening to bring the foreign trade of the kingdom to a halt, and indeed by taking steps to make good that threat, the two British banks had forced the Siamese administration into a substantially smaller revaluation of the baht than it had initially sought,[7] and had secured from the authorities an understanding that their losses arising from official alteration of the exchange rate would be minimized. It would not have escaped the attention of the administration that if there had been a Siamese bank conducting exchange business in Bangkok during this crisis, the position of the British banks would have been much weaker and, consequently, the government would have been able to maintain firm control over its exchange policy.

In fact, the prospects for the establishment of a Siamese bank had for some time been under consideration by Prince Mahit, the Minister of Finance from August 1896. At the end of the 1890s, the Prince and the British advisers in his Ministry studied a proposal for the creation of a government bank.[8] The British officials were strongly discouraging, leading Prince Mahit to suspect that they were determined to protect the interests of their own banks in the kingdom. Consequently, the Minister of Finance decided to drop the proposal for a government bank (presumably he feared that it would be difficult to exclude the administration's advisers from an official initiative of this kind), and began to think in terms of a private bank. The fact that in this same period the British Legation in Bangkok made clear to the Siamese authorities its determination to safeguard the interests of the established British banks,[9] can only have encouraged Prince Mahit to pursue his ambitions privately, and indeed in the utmost secrecy.

It is difficult to recount precisely Prince Mahit's reasons for wishing to establish a bank, or indeed the actual events surrounding its inauguration in October 1904. As the bank was a private concern, its papers did not find their way into the government records which now comprise the holdings of the National Archives in Bangkok. Indeed it is only because the bank became the focus of a diplomatic argument in 1906, and thus the subject of official correspondence, that some insight into Prince Mahit's perceptions and reasoning in 1904 can be secured. Undoubtedly, the Minister of Finance regarded the bank as a promising business venture, potentially a very profitable investment at a time when the internal production and foreign trade of the kingdom were expanding rapidly; and even if Prince Mahit himself was not primarily concerned with personal gain, his partners in the bank certainly were. The Prince's later correspondence indicated two further considerations. The first was a concern that the banking facilities available to the non-European population of the capital were severely inadequate.[10] According to the Minister, prior to his initiative the local population had to rely on pawnshops, most of which appear to have been little more than rackets run from private houses by old women which no person of rank would dare patronize. And of course pawnshops would only undertake to provide credit; they had no facilities to receive cash deposits. In these circumstances it might have been expected that the European banks in Bangkok would have hastened to provide full banking services to the local population; but, again according to Prince Mahit, those banks had a reputation for treating Siamese and Chinese customers, particularly those who did not speak English, in a very disparaging manner.[11] The second consideration was the determination of the Minister of Finance to break the British banks' domination of exchange business in Bangkok. In a letter to the King in 1906, he argued that the banks' ability to bring the trade of the port to a halt gave Britain a stunning advantage in any diplomatic–military conflict with Siam.[12] Consequently, the establishment of a Siamese bank, that would ensure the continued financing of foreign trade during an international crisis, was seen as a crucial element in the kingdom's struggle to protect its political sovereignty. Prince Mahit further argued that, through their domination of exchange operations, the foreign banks had 'squeezed the blood from our traders'.[13] A local bank would break that stranglehold—and, unlike the foreign banks, would keep its profits within the kingdom. Finally, although the Minister made no reference in this later correspondence to the coercive intervention by the Hongkong and Shanghai Bank and the Chartered Bank when Siam had abandoned silver in November 1902, that episode clearly indicated a further compelling reason to break the European banks' domination of exchange business.

Prince Mahit launched the new enterprise on 4 October 1904, using a two-storey brick building belonging to the Privy Purse Department located in the Bān Mǭ district of the capital.[14] The initial investment in the concern, at 110,000 baht, was quite modest.[15] The capital was raised primarily from within Prince Mahit's own circle of friends, and although it is not possible to identify individuals, the subsequent history of the bank would suggest that they were predominantly Sino-Siamese merchants and tax-farmers. The bank was under the management of the Minister of Finance himself, assisted by a staff of eighteen. It did not at this point undertake exchange business, but was involved only in providing deposit and loan facilities.

Preparation for the opening of the bank had been carried out unobtrusively. Prince Mahit does not appear to have sought guidance from the European officials in his Ministry, for to have done so would have made his intentions known throughout the European community in Bangkok. Rather, he relied on textbooks on bank administration and on bookkeeping to develop his knowledge of banking practices.[16] Moreover, with the enterprise actually established, Prince Mahit initially sought to confuse the public as to its true nature, most notably in his selection of a name for the bank—it was called 'Book Club'. As the Minister later explained, it was his hope that this unusual name—a transliteration of two English words which carried no meaning in Siamese—would strike the local population as droll and intriguing, and in this way would help to break down their apparent suspicion of this new form of institution and encourage them to deposit their capital in it.[17] The European community was also to be misled, although in a different direction. Europeans, the Minister argued, would simply accept that the new enterprise was a public library: thus, were the bank to fail, there would be no public disgrace, since no one would know that the concern was a bank, and it would be possible to revive the project at a later stage. It is also possible that Prince Mahit feared that, were the European banks in Bangkok to know that the new enterprise was a bank, they would make every effort to strangle it at birth.[18]

Prince Mahit's fears for the commercial survival of his bank proved to be unfounded. Indeed the 'Book Club' enjoyed a quite remarkable early success, as the following figures indicate:

The 'Book Club': Deposit and Withdrawal Transactions, October 1904 to December 1905 (in baht)

	Deposits	Withdrawals	Balance
October 1904	130,176	114,815	15,360
April 1905	1,138,800	1,002,384	438,305
December 1905	2,971,085	2,829,335	781,377

Source: Enclosed in Prince Mahit to Chulalongkorn, 19 January 1906, NA r5 KS 12.2/12.

In its first year the 'Book Club' returned a profit of 90,000 baht, on an initial investment of 110,000 baht.[19] Consequently, by early 1906 Prince Mahit had sufficient confidence to undertake major changes in the bank's managerial structure, and a substantial expansion of both its capital resources and its activities. The first element in this reorganization, initiated in February 1906, involved the sale of 3,000 shares, to expand the bank's capital by 3 million baht.[20] Of that total issue, 1,300 shares were acquired by Prince Mahit and the partners who had been involved in the founding of the 'Book Club', and a further 300 were purchased by the Privy Purse Department.[21] Of the remaining shares, 330 were acquired by the Deutsch-Asiatische Bank and 240 by the Danske Landmannsbank.[22] With the capital foundations of the bank thus secured, in April 1906 a board of directors was appointed that reflected the major shareholding interests. There were four Siamese directors, one an appointment by the Privy Purse Department, the remaining three (who were in fact Sino-Siamese) Prince Mahit's original partners in the enterprise; the board was completed by three European directors, appointed by the German and Danish banks.[23] The final element in the reorganization of the 'Book Club' was its expansion into exchange business.[24] Undoubtedly, it was this which accounted for the involvement of the Deutsch-Asiatische Bank and the Danske Landmannsbank in the local bank, for the Europeans could provide not only the overseas agencies necessary for the conduct of exchange business but also the specialist staff experienced in this aspect of banking. The importance of the European banks to the 'Book Club' was made evident by the fact that although the Deutsch-Asiatische Bank and the Danske Landmannsbank together held less than one-fifth of the bank's shares, their officials occupied three of its seven directorships, and one of them, Felix Kilian of the Deutsch-Asiatische, became manager of its foreign department.[25]

It will have been noted that Prince Mahit held no formal position in the reorganized 'Book Club'. The reason was simply that the Minister, suffering from tuberculosis, was no longer able to sustain his earlier commitment to the work of the bank: indeed, in February 1906 he was forced to take leave from his ministerial duties and seek recuperation by the sea.[26] In fact, it may well have been Prince Mahit's ill health, as much as the immediate commercial success of the 'Book Club', which prompted the bank's reorganization and expansion in the early months of 1906, for the

Minister was clearly concerned to place the business on to a more secure footing before his condition forced him to withdraw.[27] With the managerial restructure and capital expansion of the 'Book Club' completed in April 1906, it remained only to secure the bank's legal status, a particular concern here being the need to protect the shareholders from unlimited liability for the bank's debts.[28] In the absence of company legislation in Siam in this period, the legal position of the bank could be secured only by the granting of a royal charter. Prince Mahit clearly saw the drafting of this instrument as his last important responsibility towards the 'Book Club', after which he would withdraw from its affairs.[29] Towards the end of April, he informed the King that the formal petition for a royal charter was virtually complete:[30] but within a short time Prince Mahit found himself embroiled in a fierce diplomatic dispute over his involvement with the 'Book Club', a dispute which marred the end of his distinguished ministerial career and also, quite possibly, hastened his death.

The British banks in Bangkok appear to have remained unaware of the full import of the changes underway at the 'Book Club' until the managerial and capital reorganization of the Siamese concern had been largely completed. As a clearer picture began to emerge, they pressed the British Legation to make unofficial inquiries of the Siamese authorities.[31] The British Minister in Bangkok, Ralph Paget, became particularly concerned by three aspects of the case: that the Deutsch-Asiatische was closely involved in the management of the Siamese bank; that the local bank was to undertake exchange business; and that Prince Mahit had allegedly used his position as Minister of Finance to further the interests of his own bank.[32] On this last point, in mid-May 1906 he wrote to Jan Westengard, the General Adviser:

... the new Bank is favoured by unfair advantages over the other Banks by the fact that the Minister of Finance is interested in its promotion. . . . I am informed that Prince Mahit has made use of the power given him by his official position as Minister of Finance, to transfer Government balances kept by the British Banks to the charge of the new Bank and further, that H. R. H. has used his influence to induce former customers of the British Banks to desert them in favour of the new Bank.[33]

Prince Mahit strongly denied Paget's allegations, both in conversation with Westengard and in two emotionally-charged letters to the King. To Westengard he explained:

The distribution of deposits of Government funds as between, say, the Hongkong and Shanghai Banking Corporation and the Chartered depends upon the accident of which bank the money happens to be paid into by the persons who owe money to the Government. Furthermore, the Hongkong and Shanghai Banking Corporation does not pay interest on deposits above a certain sum ... and therefore the Government does not keep a large balance at that bank.[34]

In further denying that he had influenced private individuals to leave the European banks and place their accounts with the 'Book Club', Prince Mahit pointed out that many of the former banks' most important customers had become shareholders in the Siamese concern and that it was therefore only to be expected that they would transfer their business to the bank in which they had now acquired a financial interest.[35] It had not been necessary for the Minister to entice important private accounts to the 'Book Club'—they had moved of their own volition.

Prince Mahit was in no doubt as to why the allegations were being made against him. As he noted in a letter to the King in early May 1906, the British banks in Bangkok were jealously determined to keep the kingdom's exchange business to themselves, and they well understood that, left unimpeded, the 'Book Club' would rapidly pose a serious challenge to them.[36] It was not a coincidence that the storm had broken at precisely the time the 'Book Club' had begun to undertake exchange transactions. In a further letter to the King, written towards the end of May, Prince Mahit was even more explicit: 'Paget suspects that I am the principal figure—the culprit—in this matter. He believes that if he attacks me and secures my removal, this might lead the 'Book Club' to withdraw from exchange business: by unnerving me he intends that all thought of exchange banking will be abandoned.'[37]

Despite the firmness of his denials, it is clear that Prince Mahit was greatly shaken by the controversy which had erupted around him. In the first of his two letters to the King he anxiously pressed Chulalongkorn to sanction the royal charter for the 'Book Club' as soon as that instrument was prepared, arguing that without legal security, specifically the provision of limited liability, the bank's stability could be threatened by clamorous rumour (of the kind then current); indeed he openly suggested that if a royal charter were not granted, his fellow shareholders would seek to have the 'Book Club' registered abroad, almost certainly in Germany.[38] This was close to a threat. Furthermore, Prince Mahit now sought to play down his role in the past and present affairs of the 'Book Club'. Most notably he told Westengard, in a conversation later reported to both Paget and Prince Devawongse, that 'in its inception he had *assisted* the Bank to organize, because it ... had come to him for advice and help',[39] and that he had 'now sold out and disposed of whatever interest he had'.[40] The final, and most revealing, indication of Prince Mahit's agitation was the fact that, with great bitterness, he offered his resignation as Minister of Finance to the King.[41]

There are two central aspects of this affair: what were Paget's intentions in alleging that the Minister of Finance was guilty of 'a breach of official morality';[42] and was the allegation justified. With regard to the first, *pace* Prince Mahit it is doubtful whether Paget seriously sought the withdrawal of the 'Book Club' from the conduct of exchange business. The two British banks themselves, in representations to the British Legation in Bangkok, pointed out that 'they would not object to any Bank opening [in Siam] on fair competitive terms';[43] their complaint was that, in their view, the new bank had been improperly assisted by the Minister of Finance. As for Paget, his principal concern appears to have been the close involvement of the Deutsch-Asiatische Bank in the 'Book Club', and the advance of German influence in Siam which that involvement would hasten. The British Minister's concern in this respect was clearly explained in a report from Westengard to Prince Devawongse:

... the Bank, (and this means the [German] Managers of the Bank) will in time acquire a knowledge and acquaintance with the finances of the Kingdom such as cannot be obtained by the Managers of any other bank in the country. Furthermore, suppose the new Bank should fail; these German Managers would undoubtedly turn over such business as they could control to one of the great German banks ... and this German Bank would then have an immense advantage over any of the other established banks here.[44]

Whether the allegations brought by Paget against Prince Mahit were well founded must, with the available archival evidence, remain an open question. Despite the Minister's firm denials, it is possible to detect in his two letters to the King in May 1906 a suggestion that his actions had not, after all, been beyond reproach.[45] Yet even if Paget's allegations—that the Minister of Finance had used his official position to transfer government funds from the British banks to the 'Book Club' and had influenced private individuals to do the same—were true, it is doubtful whether that action alone justified the resignation of a senior government figure who had held that position with distinction for almost ten years. Indeed, this was the view of a Foreign Office official in London when he read Paget's account of the affair: 'The official morality of the Finance Minister is hardly our concern, and it is open to the Siamese Government to keep their balances where and give their business to whom they like.'[46] The fact that the King accepted Prince Mahit's resignation should not be seen as evidence of serious indiscretion, for here Chulalongkorn may well have been prompted by fear that Paget's allegations, justified or not, marked the opening of a diplomatic crisis that could be defused only if the Minister stepped down.[47] Chulalongkorn must also have been influenced in this decision by Prince Mahit's deteriorating physical condition.[48]

With the resignation of Prince Mahit the censorious interest of the British Minister in the affairs of the 'Book Club' evaporated, leaving the bank essentially unscathed by this crisis. At the beginning of July 1906, the bank's directors submitted to the government their proposed articles of association and a formal petition for the granting of a royal charter.[49] In the months which followed, Westengard led discussions with the directors, and in particular with Felix Kilian, to examine these legal instruments, as a result of which one notably significant amendment was secured. At Westengard's insistence and despite strong resistance on the part of Kilian, a clause was inserted which stipulated that not more than one-third of the bank's shares could be owned by the subjects of powers possessing extraterritorial rights in Siam.[50] It might be anticipated that this clause was introduced in order to protect the Siamese character of the bank, and indeed it is surprising that in the earlier drafting of the charter no attempt had been made to limit the degree of foreign ownership in the concern. But in fact Westengard's insistence here arose from the possibility that the bank, making advances against the security of land, could in time become a major landowner as it was required to foreclose on bad debts; if the bank were to be largely foreign-owned then that land would, in effect, also pass into foreign ownership. The charter and articles of association, finally revised, were approved by the Council of Ministers at the beginning of 1907 and on 30 January 1907 they received the King's sanction.[51] From this date the bank took the title *Baeng Sayām Kammāchon Čhamkat*, the Siam Commercial Bank Limited. Prince Mahit lived just long enough to witness this final stage in the restructuring of the 'Book Club'. He died on 15 April 1907, at the age of 44.[52]

II

For the five years or so following the establishment of the Siam Commercial Bank under royal charter, it is not possible to provide a detailed account of its activities, simply because these were not in general the subject of official correspondence. The limited evidence which is available offers a rather confused view of the bank's fortunes in those years. On the one hand there are the outward signs of secure growth. For example, there was a steady and substantial rise in the half-year net profits of the Siam Commercial Bank from 1907 right to the end of the decade;[53] and in 1908 the bank was able to invest 300,000 baht in the construction of a new head office on a very favourable commercial site in the Talātnǫi district, fronting the Čhaophrayā River and adjoining the Harbour Department.[54] But there is also evidence from this period of serious tension between the local and foreign shareholders in the bank, as well as of major mismanagement of the bank's loan business. Thus in mid-1908 the Deutsch-Asiatische Bank disposed of its shares in the Siam Commercial, selling them to the German directors of the Siamese bank.[55] According to German sources, this withdrawal was prompted in part by the bank's fear that, having opened branches in a number of major cities in the East in the mid-1900s, it had over-extended itself, and in part by its disappointment that the Siam Commercial now appeared unlikely to develop into a form of state bank as the German bank had initially hoped. However, a Siamese observer suggested that the German withdrawal arose from the fact that the charter of the Siam Commercial Bank, in limiting to one-third the proportion of shares which could be held by foreign subjects, prevented the Deutsch-Asiatische from securing a dominant position in the Siamese bank.[56] Undoubtedly, a further important consideration for the German shareholder was the allegation that the Siam Commercial Bank's loan business, under the authority of a Siamese manager, had been seriously mismanaged—specifically that major loans and overdrafts had been granted on insufficient security. According to one of the German managers, that practice had caused the bank a loss of over 400,000 baht in the late 1900s and had on several occasions brought the Siam Commercial close to collapse; that loss had been covered from the profits of the bank's exchange business, under German management, and in general the Siam Commercial had been kept solvent only through the willingness of the Deutsch-Asiatische to provide the necessary support.[57]

It was against this background of dissension and mismanagement that there emerged the figure of Joo Seng Heng who was to dominate the Siam Commercial Bank in the early 1910s. Little information is available as to his origins and early career. It would appear that he was a recent immigrant from China, for he was earlier known to Western bankers in Shanghai.[58] In the 1890s he could be found as a clerk in the Bangkok opium farm, and he became manager of that farm in the early 1900s.[59] From this position he was able to accumulate sufficient capital to establish the Joo Seng Heng Bank in about 1904.[60] The bank's business expanded rapidly, until in 1908 the concern was reorganized as the Chino-Siam Bank with locally subscribed capital of three million baht. From its establishment the Chino-Siam had a close relationship with the Siam Commercial Bank. The two banks co-operated in the conduct of exchange business, while the Chino-Siam acted as the agent for the Siam Commercial in the Sampheng district of the capital;[61] and one of the directors of the Siam Commercial, Luang Sōphonphetcharat (Kim Seng Lee), was also on the board of the Chino-Siam Bank.[62] However, the relationship appears to have been a severely unbalanced one—indeed it was later suggested that by 1910 the Chino-Siam had practically no capital and survived only on an overdraft from the Siam Commercial Bank.[63] Certainly, Joo Seng Heng's bank was financially insecure.

In the early 1910s, that financial precariousness drove Joo Seng Heng into a number of banking and business alliances in an ever more desperate struggle to maintain solvency. Towards the end of 1910, the Chino-Siam Bank merged with another recently-established local bank, the Bangkok City Bank.[64] The merger was undertaken primarily to stave off the collapse of the Bangkok City Bank, which had accumulated an uncommonly high proportion of poor loans since its establishment in August 1909, for that bank was in turn highly indebted to the Chino-Siam and would undoubtedly have brought down the latter with it.[65] But if merger removed the danger of immediate insolvency, it also burdened the Chino-Siam with the Bangkok City Bank's heavy liabilities; and the Chino-Siam was further extended as it sought to assist local rice traders, including its own major shareholders,[66] through a serious trade recession from the close of the 1900s.[67] Joo Seng Heng's second principal initiative in this period concerned the establishment of the Siam Rice Milling Company, based on the mills of a prominent rice trader, tax-farmer and business associate of his, Luang Čhitčhamnongwānit.[68] The complexity of business arrangements and personal alliances running through the company is difficult to unravel, and indeed is of only minor importance in this context; it is sufficient to note that Joo Seng Heng simply used the Siam Rice Milling Company as an instrument by which his accumulating debts might be settled. For almost three years, he drained the company of every baht of its profits in his attempt to remain solvent. When, towards the end of 1913, Luang Čhitčhamnongwānit finally told him that he could no longer exploit the Siam Rice Milling Company in this way, Joo Seng Heng's final crisis was precipitated.

During the early 1910s Joo Seng Heng had also turned his attention towards the Siam Commercial Bank. At the end of 1909, the local manager of that bank, held responsible for the recent mismanagement of the Siam Commercial's domestic loan business, resigned.[69] Rather surprisingly, the Siamese shareholders, including the Privy Purse, do not appear to have been immediately concerned to find a replacement, and consequently the German foreign manager, Herr Schwarze, soon became accustomed to exercising full authority over all the Siam Commercial Bank's transactions.[70] In early 1910, Schwarze returned to Europe on leave, handing over his position to another employee of the Deutsch-Asiatische Bank, Herr Koch. While in Europe, Schwarze persuaded the Deutsch-Asiatische to reacquire a major shareholding in the Siam Commercial, but in the process he was obliged to give an assurance that he would retain full responsibility for the foreign business of the Siamese bank and would exercise firm control over its local transactions.[71] Within a few months of the return of Schwarze to Bangkok, in November 1910, the majority Siamese shareholders, led by the Privy Purse, finally appointed a new local manager for the Siam Commercial Bank: it was Joo Seng Heng.[72]

Schwarze was seriously shaken by this appointment. It brought to mind the mismanagement of the Siam Commercial Bank's loan business by Joo Seng Heng's predecessor in the late 1900s,[73] particularly when Joo Seng Heng himself began to argue that the bank must treat the local Siamese and Chinese traders among its customers far more favourably than it had in the past;[74] it demolished the assurance which Schwarze had given to the German shareholders when he had been in Europe earlier that year, and thus threatened to provoke the Deutsch-Asiatische Bank again to withdraw;[75] and, as the Chino-Siam Bank was heavily indebted to the Siam Commercial,[76] it created a potentially serious conflict of interest for Joo Seng Heng through which the latter bank could be severely damaged. At the close of 1910, in a letter to the King and in meetings with Westengard, Schwarze sought to limit the new manager's authority;[77] some months later the German Minister in Bangkok attempted, crudely, to apply pressure on the Siamese authorities towards the same end.[78] With equal determination, Joo Seng Heng held his ground.[79] As it became clear that the ambitions of the two managers could not be reconciled, in July 1911 the Deutsch-Asiatische Bank, for the second and final time, sold out its interest in the Siam Commercial Bank.[80] This action was decisive. Schwarze remained as manager of the bank, apparently assured that he had full authority over all its transactions;[81] but in reality the balance of power within the Siam Commercial had clearly shifted to the advantage of Joo Seng Heng.

Joo Seng Heng's new-found authority was clearly demonstrated in October 1912 when, at the instigation of the Siamese manager, the Siam Commercial Bank opened a branch office in the premises of the Chino-Siam Bank in the Sampheng district of the capital.[82] The board of the Siam Commercial sanctioned this arrangement in the belief that it would facilitate a reduction in the Chino-Siam Bank's substantial overdraft with them. In practical terms the new branch would take over a portion of the outstanding loans granted by the Chino-Siam; repayment of those loans was secured against the security already deposited by the Chino-Siam with the head office of the Siam Commercial for its own overdraft. In addition, the new branch was to act as the cashier for the Chino-Siam Bank. When a customer withdrew money from the Chino-Siam, the actual cash was to come from the Siam Commercial, the account of the Chino-Siam with that bank being appropriately debited; conversely monies paid into the Chino-Siam Bank were to be credited to the Chino-Siam's account with the Siam Commercial. The Chino-Siam was forbidden to overdraw on its account with the branch office of the Siam Commercial Bank with which it shared premises.

Within 15 months of the opening of the Sampheng branch, the Chino-Siam Bank had failed and the Siam Commercial Bank faced ruin. On 11 December 1913, Joo Seng Heng paid an urgent visit to the Minister for the Capital, Čhaophrayā Yomarāt.[83] He informed the Minister that the Hongkong and Shanghai Bank was pressing the Siam Commercial for the immediate payment of 300,000 baht, but that he had been able to raise only half that sum. Unless the remaining funds could be found quickly, the Siam Commercial Bank would fail. He sought the Minister's assistance. Questioned as to the causes of this crisis, Joo Seng Heng indicated that they lay in part in the recent failure of the Kwong Yik Bank in Singapore which had seriously disrupted the important Bangkok–Singapore trade. However, he later disregarded that explanation,[84] and indeed there seems little doubt that the crisis had in fact been triggered by Luang Čhitčhamnongwānit's decision in late 1913 finally to halt Joo Seng Heng's plundering of the Siam Rice Milling Company.[85]

On the same day that he saw Čhaophrayā Yomarāt, Joo Seng Heng made out to one of the European banks in the capital a number of cheques drawn on the Chino-Siam Bank.[86] Since October 1912, when the newly-established Sampheng branch of the Siam Commercial Bank had begun to act as the cashier for the Chino-Siam, it had been the accepted practice for cheques thus drawn on the Chino-Siam Bank to be sent to the head office of the Siam Commercial for settlement. The head office would refer each cheque to its branch in Sampheng: if the Chino-Siam held sufficient funds there to cover the cheque then it would be honoured by the head office, the branch account of the Chino-Siam being appropriately debited; if there were insufficient funds, the cheque would be returned to the head office, and thence back to the Chino-Siam Bank by way of the payee. On this occasion, the Chino-Siam cheques made out by Joo Seng Heng in his capacity as a manager of that bank could not be covered by the Chino-Siam's account with the Sampheng branch. Indeed on 12 December 1913 that account, which under the October 1912 arrangements was to be kept in credit, was overdrawn by 900,000 baht.[87] Nevertheless, when these cheques reached the branch, Joo Seng Heng, now in his capacity as a manager of the Siam Commercial Bank, ordered the responsible branch official to suppress them. When the cheques were thus not returned to the head office, it was understood there that they could be covered by the Chino-Siam's account at the Sampheng branch: they were therefore honoured. In brief, by ordering the branch official to retain the cheques, Joo Seng Heng caused the head office of the Siam Commercial Bank to complete payments on behalf of the Chino-Siam Bank without the Chino-Siam's account at the Sampheng branch being debited. It would appear that Joo Seng Heng repeated this manœuvre over the three days 11–13 December 1913, extracting from the Siam Commercial Bank about one million baht.[88] It is possible that Joo Seng Heng intended to have these funds promptly repaid[89]—but before any such action was taken the discrepancies in the Siam Commercial's ledgers were discovered.[90] At a lengthy meeting of the board of directors of the Siam Commercial Bank on 14 December, Joo Seng Heng, there in his capacity as a manager of the bank, admitted that he had misappropriated the bank's funds.[91] Two days later he was brought before the courts and formally charged with embezzling one million baht from the head office of the Siam Commercial Bank and 380,000 baht (in cash) from its Sampheng branch. At the same time the Siam Commercial filed a petition of bankruptcy against the Chino-Siam Bank. On 15 December the latter bank failed to open for business.[92]

The collapse of the Chino-Siam Bank in mid-December 1913 triggered a series of failures throughout Bangkok's commercial sector. Among those directly caught in the collapse were a number of the Chino-Siam's major shareholders, for the failure revealed that the bank's share capital was only half paid up; some shareholders now found themselves liable for sums of up to 100,000 baht.[93] The collapse was more severely felt among the Sino-Siamese merchants in Bangkok, for this community included the bank's principal customers.[94] On the day following the failure of the Chino-Siam Bank, Chaophraya Yomarāt reported that 34 Bangkok rice-mills had already stopped buying paddy, this leaving just 9 mills still working.[95] At the same time the *Bangkok Times* was carrying a report that two Chinese import–export firms and one rice-mill had failed, and that a further 20 mills were said to be at risk.[96] Towards the end of December 1913, a large group of Chinese merchants petitioned the King, arguing that they had been brought to the edge of ruin by the loss of those funds which had been deposited in the Chino-Siam Bank when it failed, and by the reluctance of the other banks in the capital to finance trade while the rice market remained unsettled.[97]

By far the largest rice firm threatened by the collapse of the Chino-Siam Bank, a firm which reportedly had accounted for about 40 per cent of the kingdom's rice exports during that year,[98] was the Siam Rice Milling Company. Arguing that it had held a large credit balance in the Chino-Siam when the bank had failed, the company urgently approached the four remaining Bangkok banks for an advance of 400,000 baht to secure its position.[99] But the approach was rejected, for each of the banks already had a large volume of outstanding unsecured loans with the Siam Rice Milling Company and were therefore understandably unwilling to extend further credit without clear assurance that their advances would be repaid. The government now intervened, concerned that the failure of the company at that point could well provoke a devastating collapse throughout the capital's commercial sector. Under the direction of the Minister of Finance, Prince Čhanthaburī, an arrangement was negotiated by which the Siam Commercial Bank settled the Siam Rice Milling Company's outstanding liabilities with respect to the three European banks in Bangkok, Luang Čhitčhamnongwānit undertaking to repay the Siam Commercial as the financial position of the company recovered.[100] But it very soon became clear that the Siam Rice Milling Company was broken. No payments were received from Luang Čhitčhamnongwānit from the middle of March 1914. Consequently in early May, the Siam Commercial Bank, almost certainly acting with the prior knowledge of the government and quite possibly on the administration's instructions, filed a bankruptcy petition against Luang Čhitčhamnongwānit and the Siam Rice Milling Company.[101] The latter were declared bankrupt by the civil court towards the end of June.[102]

When the Chino-Siam Bank had collapsed in mid-December 1913, the Siam Commercial Bank had come under intense pressure. On 15 December, over 400,000 baht was withdrawn from the latter bank; over the following three days a further 800,000 baht was withdrawn.[103] These withdrawals were so serious that there is little doubt that the Siam Commercial Bank too would have failed if the government had not swiftly intervened.[104] As soon as the administration had become aware of the impending crisis (that is, with the visit of Joo Seng Heng to Čhaophrayā Yomarāt on 11 December), the King had instructed that the funds of the Privy Purse and of the Treasury be used to protect the Siam Commercial.[105] The bank could not be allowed to fail. When the crisis actually broke, only the Treasury reserves were mobilized,[106] for it would appear that the Privy Purse reserves were too depleted to be of value. Prince Čhanthaburī, charged with the rescue, now justified the defence of the Siam Commercial Bank with government funds primarily in terms of the need to protect the substantial Privy Purse interests in the bank, and of the need to prevent a collapse of trade.[107]

But as the Minister of Finance well realized, a swiftly delivered injection of Treasury funds could not by itself restore financial stability or commercial credibility to the Siam Commercial Bank. It was clear that the administrative practices of the bank were very seriously flawed (indeed, he argued, this had been clear from the time of Joo Seng Heng's appointment to the Siam Commercial);[108] and there was also the emerging probability that, given the severity of the commercial crisis which had been triggered by the Chino-Siam failure, the Siam Commercial Bank would have to be financially restructured. From the time the crisis broke, Prince Čhanthaburī and his Ministry assumed principal responsibility for carrying through that reconstruction, although clearly the bank was a private institution.

That responsibility occupied the Ministry of Finance throughout the opening months of 1914. The first task was to secure an accurate statement of the Siam Commercial Bank's losses arising from the Chino-Siam collapse. Initially, these had been placed at around 2 million baht, but a detailed examination of the bank's accounts undertaken by a European official from the Ministry of Finance revealed that the Siam Commercial had lost no less than 5,747,000 baht.[109] These losses had been incurred in part directly through the failure of the Chino-Siam Bank and the embezzlement committed by Joo Seng Heng, but also indirectly through the subsequent failure of a number of prominent merchants indebted to the Siam Commercial and through the discovery that the bank had earlier purchased a substantial volume of forged commercial bills. Losses on that scale not only wiped out the bank's original capital issue (3 million baht) and its accumulated reserves (1,137,906 baht) but also absorbed virtually all the funds which the Ministry of Finance had made available to it from mid-December 1913—funds which clearly it had been intended the bank would in time repay. Indeed, in the aftermath of the Chino-Siam crisis, the Siam Commercial Bank's only remaining assets were its buildings and land holdings, valued together at just 300,000 baht.

After considering these sombre details, Prince Čhanthaburī proposed to the King that the Siam Commercial Bank be financially reconstructed through a new 3 million baht share issue, all save 300,000 baht of the bank's existing capital issue being written off.[110] The King approved.[111] Only then were the proposals reported to the board of directors of the bank. They were subsequently brought to two extraordinary general meetings of the bank's shareholders, held in May and June 1914.[112] Both meetings were stormy occasions, for a minority of shareholders (led by Europeans in the government's employ) were angered by the disclosure that the bank had lost heavily through its purchase of forged commercial bills and its granting of questionable loans which had collapsed during the Chino-Siam crisis. That minority pressed the government not to approve the reconstruction proposals;[113] but as the proposals had originated from the government itself and as the government élite commanded a majority of the shares, their complaint could be dismissed. The new share issue was successfully floated a few weeks later. The Privy Purse invested 1,634,000 baht in the issue,[114] and thus maintained a majority interest in the Siam Commercial Bank. It might be added that the Privy Purse was able to make this major investment only with the assistance of an equivalent loan from the Treasury. Prince Čhanthaburī approved the use of government funds in this manner with the argument that the investment would make secure the re-establishment of the Siam Commercial Bank which in itself was 'for the common good'.[115]

During the same opening months of 1914 as Prince Čhanthaburī was preparing the reconstruction of the Siam Commercial, Joo Seng Heng stood before the Criminal Court charged with embezzling 1,300,000 baht from that bank.[116] Joo Seng Heng did not challenge the prosecution's account of the events of the previous December, but he strenuously maintained that he 'had never taken a single tical for his own use'.[117] This was a distressingly weak defence, and in early April 1914, Joo Seng Heng, found guilty as charged, was sentenced to 10 years' imprisonment and fined 10,000 baht.[118]

As Joo Seng Heng was sentenced, the editor of the *Bangkok Times* drew his readers' attention to what was by far the most disturbing aspect of the recent crisis.

Nai Chalong [Joo Seng Heng] ought not to have been entrusted with the power and position that were his.... Those who recommended and appointed him must have known that in no other country in the world would a man have been selected for the position of manager of a responsible Bank after having served a sentence of imprisonment for misappropriation of money.... Practically he was the manager, not of one

Bank merely, but of two, and one of them, largely his own property, known to be none too sound.[119]

In fact Joo Seng Heng's appointment had been criticized at the time, on the grounds that he had very substantial interests in the Chino-Siam Bank.[120] Moreover, Prince Čhanthaburī was later to claim that from the moment of Joo Seng Heng's appointment to the Siam Commercial Bank he had seen that a major collapse was inevitable.[121] But these observations simply make the appointment all the more disturbing. Why indeed was Joo Seng Heng appointed?

It is tempting here to point to naïvety or incompetence on the part of the board of directors and major shareholders of the Siam Commercial Bank; but this is unlikely to provide a sufficiently full explanation. Two other, related, influences may have been important. The first was the determination of the bank's majority Siamese shareholders in the early 1910s to curb the European domination of the bank's management by the appointment of a powerful local manager;[122] the second was the acute shortage of experienced local banking entrepreneurs. Thus the Siam Commercial Bank's principal Siamese shareholders, including notably the Privy Purse, may have been sufficiently impressed with Joo Seng Heng's banking experience (after all, he was the founder and major shareholder in the Chino-Siam Bank), to wish to overlook the disreputable elements in his background and the fact that his appointment to the Siam Commercial could well create a serious conflict with his existing interests. They took a risk; and they lost.

III

For the period following the successful reconstruction of the Siam Commercial in mid-1914, the official documentation concerning the bank now held in the National Archives in Bangkok is very slight. It is thus possible to make just two broad points about the later history of the bank. First, the bank's principal Siamese shareholders, perhaps unnerved by their disastrous experience with Joo Seng Heng, left the management of the Siam Commercial in European hands. Herr Willeke, manager during the Chino-Siam crisis, remained at the bank until July 1917 when, with Siam's declaration of war against the Central Powers, he was removed by the Siamese authorities.[123] For his successor the Siam Commercial Bank secured the services of Mr G. H. Ardron, then the accountant

THE EARLY DEVELOPMENT OF INDIGENOUS BANKING

at the Bangkok branch of the Hongkong and Shanghai Bank. He remained with the Siamese bank until his retirement in 1927. Ardron was followed through the 1930s by a succession of British and American bankers,[124] so that at the outbreak of war in December 1941 the Siam Commercial's manager was an American, W. K. Le Count, assisted by an Englishman, C. W. Taylor.[125] It was only with their internment that foreign management of the bank came to an end.[126] Second, in this period the Siamese administration remained prepared to provide crucial assistance to the Siam Commercial Bank during times of crisis. Thus, during the world depression of the early 1930s the King instructed the Ministry of Finance to keep its withdrawals from the bank to a minimum and, if possible, to increase its deposits in order to secure the bank's position:[127] and when the baht value of the bank's sterling balances and investments was severely cut by Britain's abandonment of gold and devaluation in late 1931, the Privy Purse quickly came to the aid of the Siam Commercial with a 3 million baht interest-free loan.[128] Such intervention, over and above the common preference shown to the bank in the distribution of government financial business,[129] was central to the Siam Commercial Bank's continued success. This is an argument to which I will return in the concluding chapter.

    1. Maurice Collis, *Wayfoong: The Hongkong and Shanghai Banking Corporation*, London, 1965, p. 90; Compton Mackenzie, *Realms of Silver: One Hundred Years of Banking in the East*, London, 1954, p. 195.
    2. Paul Sithi-Amnuai, *Finance and Banking in Thailand: A Study of the Commercial System, 1888–1963*, Bangkok, 1964, p. 33.
    3. Collis, op. cit., p. 91.
    4. Thiravet Pramuanratkarn, 'The Hongkong Bank in Thailand: A Case of a Pioneering Bank', in Frank H. H. King (ed.), *Eastern Banking: Essays in the History of the Hongkong and Shanghai Banking Corporation*, London, 1983, p. 424.
    5. Mackenzie, op. cit., p. 196; Collis, op. cit., p. 92.
    6. The following draws on Ian Brown, 'Siam and the Gold Standard, 1902–1908', *Journal of Southeast Asian Studies*, vol. 10, no. 2 (September 1979), pp. 383–5.
    7. In fact, just as the government and the British banks reached formal agreement, the former began a gradual revaluation of the baht in response to a rise in the gold-price of silver, so that by September 1903 it had reached the rate against sterling originally announced in November 1902 (Brown, 'Siam and the Gold Standard, 1902–1908', p. 386).

8. Prince Mahit to Chulalongkorn, 19 January 1906, NA r5 KS 12.2/12.

9. On 10 November 1899, the *Bangkok Times* carried a report that the Danske Landmannsbank of Copenhagen, in alliance with French, Russian, and German banks, had secured a concession for 'the National Bank of Siam'. The British Minister, George Greville, immediately wrote to Prince Devawongse (11 November 1899, NA r5 Kh 20/4) asking for a denial of the report, but adding that if it were true 'it would seriously interfere with vested British interest and might lead to complications'. In his reply (11 November 1899, NA r5 Kh 20/4) the Prince dismissed the report as 'a perfect canard'.

10. Prince Mahit to Chulalongkorn, 19 January 1906, NA r5 KS 12.2/12.

11. Even in the early 1920s the European banks declined 'to receive money on deposit from any person who is unacquainted with some European language' (W. A. Graham, *Siam*, London, 1924, vol. 1, p. 341). This provision, of course, would not have prevented them from securing the valuable custom of the Siamese élite.

12. Prince Mahit to Chulalongkorn, 24 May 1906, NA r5 KS 12.2/12.

13. Prince Mahit to Chulalongkorn, 9 May 1906, NA r5 KS 12.2/12.

14. Siam Commercial Bank, *Thīralu'k wanpōet samnakngānyai thanākhānthaiphānit čhamkat 19 singhākhom 2514* (*To Commemorate the Opening of a New Head Office of the Siam Commercial Bank, 19 August 1971*), Bangkok, 1971, p. 28.

15. Prince Mahit to Chulalongkorn, 19 January 1906, NA r5 KS 12.2/12.

16. Ibid.

17. Ibid.

18. Prince Mahit's ruse appears to have been only partly successful. On 3 October 1904, the day prior to the formal opening of the 'Book Club', the *Bangkok Times* reported that it was 'a general commission agency, a bank and other things. It is to hold a public auction of jewellery on Saturday.' In one minor respect the Prince's choice of name caused spectacular confusion. On 16 February 1906 that same newspaper informed its readers that 'Book Club' was a corruption of a Pali word.

19. Prince Mahit to Chulalongkorn, 19 January 1906, NA r5 KS 12.2/12.

20. Ibid.; Prince Mahit to Chulalongkorn, 20 February 1906, NA r5 KS 12.2/12.

21. Prince Mahit to Chulalongkorn, 19 January 1906; Chulalongkorn to Prince Sommot, 24 February 1906, NA r5 KS 12.2/12.

22. Siam Commercial Bank, op. cit., p. 29.

23. Ibid.; Prince Sommot to Chulalongkorn, 26 April 1906, NA r5 KS 12.2/12.

24. Prince Mahit to Chulalongkorn, 19 January 1906; 20 April 1906, NA r5 KS 12.2/12.

25. Joo Seng Heng to Vajiravudh, 18 July 1911, NA r6 Kh 15/2.

26. Prince Mahit to Chulalongkorn, 20 February 1906, NA r5 KS 12.2/12.

27. At the beginning of the year, as the reorganization of the 'Book Club' was being discussed, Prince Mahit told the King that, during his illness, he feared that were he to die, the bank would surely fail. (Prince Mahit to Chulalongkorn, 19 January 1906, NA r5 KS 12.2/12.)

28. Prince Mahit to Chulalongkorn, 19 January 1906, NA r5 KS 12.2/12.

29. Ibid.; Prince Mahit to Chulalongkorn, 20 February 1906, NA r5 KS 12.2/12.

30. Prince Mahit to Chulalongkorn, 20 April 1906, NA r5 KS 12.2/12.

THE EARLY DEVELOPMENT OF INDIGENOUS BANKING   147

31. Westengard to Prince Devawongse, 16 May 1906, NA r5 KS 12.2/12.
32. Paget to Sir Edward Grey, 15 May 1906, PRO FO 371/132.
33. Paget to Westengard, 15 May 1906, NA r5 KS 12.2/12.
34. Westengard to Prince Devawongse, 22 May 1906, NA r5 KS 12.2/12.
35. Ibid.
36. Prince Mahit to Chulalongkorn, 9 May 1906, NA r5 KS 12.2/12.
37. Prince Mahit to Chulalongkorn, 24 May 1906, NA r5 KS 12.2/12.
38. Prince Mahit to Chulalongkorn, 9 May 1906, NA r5 KS 12.2/12.
39. Westengard to Paget, 1 June 1906, PRO FO 371/132. Emphasis added.
40. Westengard to Prince Devawongse, 22 May 1906, NA r5 KS 12.2/12.
41. Prince Mahit to Chulalongkorn, 9 May 1906 and 24 May 1906, NA r5 KS 12.2/12.
42. Paget to Sir Edward Grey, 14 May 1906 (Telegram), PRO FO 371/132.
43. Paget to Sir Edward Grey, 15 May 1906, PRO FO 371/132.
44. Westengard to Prince Devawongse, 22 May 1906, NA r5 KS 12.2/12. See also, Paget to Sir Edward Grey, 15 May 1906, PRO FO 371/132.
45. There is a further hint of impropriety in an earlier letter from Prince Mahit to Chulalongkorn (20 February 1906, NA r5 KS 12.2/12): 'A large number of Chinese and Siamese customers are almost certain to use the 'Book Club' because we have ways to herd them in ...'.
46. Minute on Paget to Sir Edward Grey, 15 May 1906, PRO FO 371/132.
47. The Siamese administration was surprised (and surely disturbed) to learn that the affair had been discussed between the British Foreign Secretary, Sir Edward Grey, and the French Ambassador in London. (Westengard to Prince Devawongse, 22 May 1906, NA r5 KS 12.2/12.) It might also be noted that when Prince Mahit resigned, Westengard informed Paget (1 June 1906, PRO FO 371/132) that he had done so 'to shield the Government from any annoyance or difficulty'.
48. The official announcement of the Minister's resignation explained that it was on the grounds of ill health. (*Bangkok Times*, 6 June 1906.) The King noted briefly: 'Prince Mahit is ill and has resigned; I think that this problem has been halted.' (Chulalongkorn to Prince Devawongse, 30 May 1906, NA r5 KS 12.2/12.)
49. Čhaophrayā Thēwēt to Chulalongkorn, 7 July 1906, NA r5 KS 12.2/12.
50. Westengard, 'On a Proposed Charter for Siam Commercial Bank Limited', 21 July 1906; Westengard to Čhaophrayā Thēwēt, 8 October 1906, NA r5 KS 12.2/12.
51. Chulalongkorn to Čhaophrayā Thēwēt, 30 January 1907, NA r5 KS 12.2/12.
52. *Bangkok Times*, 16 April 1907.
53. Papers from the Office of the Financial Adviser. (When these papers were held in the Library of the Ministry of Finance, the reference here was 26/E. This section of the Adviser's records does not appear to be included in the National Archives' holdings.)
54. Siam Commercial Bank, op. cit., p. 31; Siam Commercial Bank, Report of the Board of Directors to the Fourth General Meeting, 20 October 1908, NA r5 KS 12.2/12.
55. Frank H. H. King, 'The Foreign Exchange Banks in Siam, 1888–1918, and the National Bank Question', unpublished paper presented to the International Conference on Thai Studies, Bangkok, August 1984, pp. 6–7.
56. Joo Seng Heng to Vajiravudh, 18 July 1911, NA r6 Kh 15/2.
57. Schwarze to Vajiravudh, 14 December 1910, NA r6 Kh 15/2. In making

these observations, Schwarze did not imply fraudulent practice on the part of the Siamese manager; and, indeed, in commenting on this letter, the Financial Adviser explained that the manager responsible for granting loans to local residents inevitably had to contend with personal and social pressures which could lead him, against his better judgement, 'to agree to transactions involving undue risk of loss to the Bank'. (Williamson, 'Memorandum by the Financial Adviser on the petition presented to His Majesty by Mr. Schwarze, Foreign Manager of the Siam Commercial Bank', 23 December 1910, NA r6 Kh 15/2.) It must be added that Prince Damrong, in contrast, suspected some measure of irregularity. (Meeting of the Council of Ministers, 26 December 1910, NA r6 Kh 15/2.)

58. *Bangkok Times*, 14 March 1914.
59. *Bangkok Times*, 4 March 1914.
60. Ibid.; G. William Skinner, *Chinese Society in Thailand: An Analytical History*, Ithaca, 1957, p. 109.
61. Siam Commercial Bank, Report of the Board of Directors to the Fourth General Meeting, 20 October 1908, NA r5 KS 12.2/12.
62. Siam Commercial Bank, op. cit., p. 29; Notification announcing the establishment of the Chino-Siam Bank, 21 February 1908, NA r5 Kh 20/8.
63. *Bangkok Times*, 7 April 1914.
64. Čhaophrayā Wongsānupraphat to Vajiravudh, 25 November 1910, NA r6 Kh 15/1.
65. *Bangkok Times*, 14 March 1914; Phrayā Ratsadākǫnkōson to Vajiravudh, December 1913, NA r6 Kh 15/3; Board of the Chino-Siam Bank to Čhaophrayā Yomarāt, 5 June 1911, NA r6 Kh 15/1.
66. Skinner, op. cit., p. 158.
67. Board of the Chino-Siam Bank to Čhaophrayā Yomarāt, 5 June 1911, NA r6 Kh 15/1.
68. Phrayā Ratsadākǫnkōson to Vajiravudh, December 1913, NA r6 Kh 15/3; Luang Čhitčhamnongwānit to Vajiravudh, 7 May 1914, NA r6 Kh 15/6; *Bangkok Times*, 23 July 1914.
69. Williamson, 'Memorandum by the Financial Adviser on the petition presented to His Majesty by Mr. Schwarze, Foreign Manager of the Siam Commercial Bank', 23 December 1910, NA r6 Kh 15/2; Meeting of the Council of Ministers, 26 December 1910, NA r6 Kh 15/2.
70. Joo Seng Heng to Vajiravudh, 18 July 1911, NA r6 Kh 15/2.
71. Ibid.; Schwarze to Vajiravudh, 14 December 1910, NA r6 Kh 15/2.
72. Von der Goltz (German Minister in Bangkok) to Prince Devawongse, 5 July 1911; Joo Seng Heng to Vajiravudh, 18 July 1911, NA r6 Kh 15/2. It was after this appointment that Joo Seng Heng removed his queue and received the title, Nāi Chalǫng Naiyanāt (*Bangkok Times*, 15 December 1913). In the interests of clarity, this account will continue to use his Chinese name.
73. Schwarze to Vajiravudh, 14 December 1910, NA r6 Kh 15/2.
74. Joo Seng Heng to Vajiravudh, 11 June 1911, NA r6 Kh 15/3.
75. Schwarze to Vajiravudh, 14 December 1910; Westengard to Prince Devawongse, 23 December 1910, NA r6 Kh 15/2.
76. *Bangkok Times*, 7 April 1914.
77. Schwarze to Vajiravudh, 14 December 1910; Westengard to Prince Devawongse, 23 December 1910, NA r6 Kh 15/2.
78. Von der Goltz to Prince Devawongse, 5 July 1911; Westengard to Prince

THE EARLY DEVELOPMENT OF INDIGENOUS BANKING   149

Devawongse, 6 July 1911, NA r6 Kh 15/2.
   79. Westengard to Prince Devawongse, 6 July 1911, NA r6 Kh 15/2.
   80. *Bangkok Times*, 15 July 1911.
   81. *Bangkok Times*, 24 December 1913.
   82. Siam Commercial Bank, Meeting of the Board of Directors, 7 October 1912, NA r6 Kh 15/3; *Bangkok Times*, 24 December 1913, 4 March 1914, 5 March 1914.
   83. Čhaophrayā Yomarāt to Vajiravudh, 12 December 1913, NA r6 Kh 15/3.
   84. *Bangkok Times*, 20 March 1914.
   85. Phrayā Ratsadākǫnkōson to Vajiravudh, December 1913, NA r6 Kh 15/3.
   86. Prince Čhanthaburī to Vajiravudh, 15 December 1913, NA r6 Kh 15/3.
   87. *Bangkok Times*, 4 March 1914.
   88. Prince Čhanthaburī to Vajiravudh, 15 December 1913, NA r6 Kh 15/3.
   89. This point was raised by the *Bangkok Times* as it first reported the crisis (15 December 1913); and Joo Seng Heng himself was soon to argue that he had 'fully intended that the money should be paid back' (*Bangkok Times*, 20 December 1913). Indeed, since Joo Seng Heng must have been aware that the discrepancies in the Siam Commercial accounts would rapidly come to light, there is some reason to believe that this was in fact his intention.
   90. *Bangkok Times*, 15 December 1913.
   91. *Bangkok Times*, 15 December 1913, 17 December 1913.
   92. *Bangkok Times*, 15 December 1913, 16 December 1913.
   93. Phrayā Ratsadākǫnkōson to Vajiravudh, December 1913, NA r6 Kh 15/3; *Bangkok Times*, 19 December 1913.
   94. *Bangkok Times*, 18 December 1913.
   95. Čhaophrayā Yomarāt to Vajiravudh, 16 December 1913, NA r6 Kh 15/3.
   96. *Bangkok Times*, 16 December 1913.
   97. Petition from 136 Chinese merchants to Vajiravudh, 27 December 1913, NA r6 Kh 15/3.
   98. The managers of the four Bangkok banks to Čhaophrayā Yomarāt, 17 December 1913, NA r6 Kh 15/3.
   99. Ibid.
   100. Prince Čhanthaburī to Vajiravudh, 18 December 1913, NA r6 Kh 15/3; *Bangkok Times*, 9 June 1914.
   101. *Bangkok Times*, 11 May 1914, 9 June 1914.
   102. *Bangkok Times*, 23 June 1914.
   103. *Bangkok Times*, 16 December 1913, 19 December 1913.
   104. The managers of the four Bangkok banks to Čhaophrayā Yomarāt, 17 December 1913, NA r6 Kh 15/3.
   105. Vajiravudh to Čhaophrayā Yomarāt, 12 December 1913, NA r6 Kh 15/3.
   106. Prince Čhanthaburī to Vajiravudh, 15 December 1913, NA r6 Kh 15/3.
   107. Ibid.
   108. Ibid.
   109. Prince Čhanthaburī to Vajiravudh, 13 April 1914, NA r6 Kh 15/2; *Bangkok Times*, 8 May 1914.
   110. Prince Čhanthaburī to Vajiravudh, 13 April 1914, NA r6 Kh 15/2.
   111. Vajiravudh to Prince Čhanthaburī, 25 April 1914, NA r6 Kh 15/2.
   112. Prince Rātburī to Vajiravudh, 13 June 1914, NA r6 Kh 15/2; *Bangkok Times*, 25 May 1914, 11 June 1914.
   113. *Bangkok Times*, 4 July 1914.

114. Official in the Privy Purse Department to Vajiravudh, 13 August 1914, NA r6 Kh 15/2.

115. Prince Čhanthaburī to Vajiravudh, 1 May 1914, NA r6 Kh 15/2.

116. The trial was reported in detail in the *Bangkok Times* over the period February–April 1914.

117. *Bangkok Times*, 23 March 1914. See also, 'Testimony of Nai Chalong Naiyanath', in Chatthip Nartsupha, Suthy Prasartset, and Montri Chenvidyakarn (eds.), *The Political Economy of Siam 1910–1932*, Bangkok, 1981, pp. 159–78.

118. *Bangkok Times*, 11 April 1914. Also convicted on this charge, although given a lighter sentence, was Joo Seng Heng's younger brother who had been a manager of the Chino-Siam Bank.

119. *Bangkok Times*, 11 April 1914. No further reference to Joo Seng Heng's earlier conviction for misappropriation of money has been found.

120. *Bangkok Times*, 16 December 1913.

121. Prince Čhanthaburī to Vajiravudh, 15 December 1913, NA r6 Kh 15/3.

122. When the conflict between Joo Seng Heng and Schwarze erupted at the close of 1910, the Siamese shareholders argued that as four-fifths of the bank's capital was in Siamese hands, by right the bank should be run by Siamese. In fact on this occasion they were seeking only equal authority as between the local and foreign manager (Westengard to Prince Devawongse, 23 December 1910, NA r6 Kh 15/2).

123. *Bangkok Times*, 23 July 1917; King, op. cit., pp. 16–17.

124. *Bangkok Times*, 2 February 1932; Virginia Thompson, *Thailand: The New Siam*, New York, 1941, p. 592.

125. Siam Commercial Bank, op. cit., pp. 35–6.

126. Sithi-Amnuai, op. cit., p. 35.

127. Meeting of the Supreme Council, 7 June 1932, NA r7 Kh 10/1.

128. Siam Commercial Bank, op. cit., pp. 34–5.

129. By the early 1930s almost all government funds were handled by the Siam Commercial Bank (Thiravet, op. cit., p. 430).

5
The State and the Promotion of Economic Diversification

THE Siam Commercial Bank, in decisively breaking the European banks' monopoly of modern banking in the kingdom, constituted an important diversification in the commercial structure of Siam. This chapter explores a number of important forms of economic diversification which also were attempted in the kingdom in the early twentieth century. In the promotion of each, the administrative élite played the central role, although its initiative took a variety of significantly different forms. The opening section of the chapter considers two business enterprises—the Siam Cement Company which, like the Siam Commercial Bank, involved a private initiative on the part of the élite; and the Siamese Steamship Company, which was a part-government venture. The closing section considers attempts by the Siamese administration in the early twentieth century to encourage the emergence of a technologically more advanced silk industry, and to promote a major expansion of cotton cultivation in the kingdom.

I

In 1908 a group of European and Siamese entrepreneurs, including Louis T. Leonowens (the son of Anna Leonowens and the owner of a major teak extraction firm), proposed the formation of a company to undertake the manufacture of cement in Siam.[1] The prospects for the new venture were very encouraging. Most importantly, there existed a substantial and rising local demand for cement, a demand created by the government's various public works projects (notably its railway construction programme) and by the large volume of private construction work being undertaken, in particular, in the capital itself. That demand could be met, of course, only by imports. At the end of the 1900s cement imports averaged almost 16,000 tons per annum, Siam's principal suppliers being French Indo-China (there was a major plant at Haiphong), Denmark, and Hong Kong.² However, imported cement was expensive, partly because it had to bear high freight charges,³ but also, it was argued, because the Bangkok cement dealers used their domination of this market to force up prices.⁴ Certainly, the group of local entrepreneurs was confident that it could manufacture cement for a substantially lower retail price than that currently being demanded for imported brands.⁵ Finally, as the group's investigations soon confirmed, the kingdom possessed in commercial quantities, high quality deposits of the principal raw materials required for the manufacture of cement.⁶ Limestone could be secured from Chǫng Khāe in Lopburī province, a little over 100 miles from Bangkok and very close to the northern railway; clay would be extracted at Bāng Sū', some 5 miles from the capital and again alongside the northern railway. The new company proposed to erect its factory at the Bāng Sū' site.

In July 1909, the board of directors approached King Chulalongkorn to ask whether he would take up part of a proposed major share issue.⁷ The King was excited by the prospect of establishing cement production in the kingdom,⁸ and was encouraged in this by an accompanying report by a European official from the Ministry of Public Works, E. G. Gollo, which strongly confirmed the industry's potential.⁹ However, Chulalongkorn wished for much fuller information on this particular venture before he committed himself,¹⁰ and this suggests (the documentary record runs dry at this point) that he may subsequently have declined to acquire an interest in the company. It is clear only that the company later 'failed from lack of financial support'.¹¹

Despite its premature failure, the company, through its surveys and reports, had served to draw attention to an important business opportunity, and within a short time its initiative had been taken up by a senior member of the government, Čhaophrayā Yomarāt, the Minister for the Capital. The preparatory work was now undertaken largely by E. G. Gollo. During his leave in Europe in 1912, at the request of Čhaophrayā Yomarāt, Gollo arranged for samples of Chǫng Khāe limestone and Bāng Sū' clay to be laboratory tested, and he visited Denmark to examine cement-manufacturing plant.¹² It was during this time that the initial contacts were made which were to lead, in early 1913, to preliminary discussions between Čhaophrayā Yomarāt and the Danish East Asiatic Company for the formation of a joint-venture to undertake the manufacture of cement in Siam. The initiative here was taken by the East Asiatic Company, but the discussions soon broke down over its demand that the company be at least an equal, and preferably a majority, shareholder in the proposed enterprise. With this breakdown, Čhaophrayā Yomarāt became concerned that the Danes would now proceed with the project alone, and in so doing would secure control over the limestone deposits at Chǫng Khāe, apparently the only substantial deposits in the kingdom. He therefore rapidly advanced his own plans.

The capital required to establish the cement company was estimated by Čhaophrayā Yomarāt at one million baht. The Minister found it impossible to raise this sum from within the Siamese official and commercial communities in Bangkok, and he was therefore forced to turn, in possibly similar circumstances to those of the earlier company, to the Privy Purse.[13] King Vajiravudh promptly undertook to invest 500,000 baht in the proposed cement company. With that major commitment secure, a further 250,000 baht was allocated to Luang Sawatdīwiangchai, an associate of Čhaophrayā Yomarāt, in anticipation that that block of shares would subsequently be largely taken up by his colleagues and acquaintances. In the event, only a very small proportion (300 shares out of 2,500) was taken up in this way. As Luang Sawatdīwiangchai did not have the financial resources to meet the calls on the 2,200 shares left in his name, in April 1914 Čhaophrayā Yomarāt was forced to ask the King to provide the funds to cover those payments, ownership of the shares subsequently being divided between Luang Sawatdīwiangchai and Čhaophrayā Yomarāt, although mortgaged to the Privy Purse against its loan.[14] In brief, of the one million baht required to establish the cement company, almost three-quarters was met by the Privy Purse, 500,000 baht as its own investment and 220,000 baht to secure the interests of Čhaophrayā Yomarāt and Luang Sawatdīwiangchai. The remaining 250,000 baht was provided by a Danish businessman long resident in Bangkok, Captain W. L. Grut.[15] For many years, Grut had been manager of the Siam Electric Company, but now, having recently withdrawn from that concern, he was sought by Čhaophrayā Yomarāt to be manager of the new cement company.[16]

The Siam Cement Company was formed in mid-1913 and incorporated under royal charter in December of that year.[17] Its first manager was not Captain Grut as Čhaophrayā Yomarāt had at first wished but another Dane, Oscar Schultz; however, Grut was clearly a dominant influence within the company, being a member of the board of directors from its creation and then the chairman of the board from the early 1920s to the beginning of the 1940s. The technical administration of the company was primarily in Danish hands, the Siamese concern having a close relationship here with the Danish firm of F. L. Smidth, which also supplied the major part of its plant and machinery.[18] The Siam Cement Company began production on its Bāng Sū' site in May 1915.[19]

The Siam Cement Company was a notable commercial success. Sales in 1915 were approximately 5,000 tons but rose rapidly thereafter (assisted initially by the war-time dislocation of trade which kept foreign cement out of the Siamese market) to reach 25,000 tons in 1920.[20] By 1930 the company's sales stood at approximately 70,000 tons, and in 1940, some 135,000 tons. In contrast cement imports, which had averaged a little over 22,100 tons per annum in the three years 1911/12–1913/14, fell to an annual average of 7,700 tons in the years 1920/1–1922/3.[21] In the remaining years of the inter-war decades, there were occasional substantial increases in cement imports; but at no point in this period did imports take more than one-fifth of the Siamese market and in most years the proportion was very much less. In addition to securing a domination of the local market, during the 1920s the Siam Cement Company developed a small export trade, principally to the Federated Malay States, although this trade sharply declined in the depression years of the 1930s.[22] The dramatic increase in the company's production in this period required a major expansion of the original plant. The first extension of the Bāng Sū' factory, involving the installation of a second and larger rotary kiln with corresponding grinding mills and completed in August 1923, more than doubled the plant's original capacity;[23] at the end of the 1920s additional machinery was ordered which practically doubled its capacity once more; and on the eve of the Pacific War, the company decided to construct a second factory in Saraburī district, close to a recently developed limestone deposit.[24] This repeated expansion in production capacity was largely financed by further share issues, increasing the company's share capital from the original one million baht in 1913 to four million baht in 1940.[25] The Privy Purse maintained its major interest in the company through each share issue.[26] And finally, it can be noted that from the time it began production right through the inter-war years, the Siam Cement Company secured a very substantial return on its operations. In the 1920s, the annual net profit was on average a little over 375,000 baht, and the company was able to pay a dividend of either 10 per cent or 11 per cent in each of those years.[27] Profits fell somewhat in the 1930s, but the lowest dividend paid in that decade was still 7 per cent. Thus, the Siam Cement Company provided a substantial income for its principal shareholders. As early as 1918, Čhaophrayā Yomarāt, holding almost one-quarter of the company's shares, received a dividend payment of 68,000 baht.[28] Indeed, as the Minister noted in a letter to the King, that income ensured that he could maintain his social position as a high-ranking official; he would not be able to do so if he relied solely on his government salary.[29]

It is not difficult to account for the success of this important industrial initiative. As noted above, there existed in Siam in this period, a substantial and rising local demand for cement—certainly a sufficient demand to absorb the full output of a major plant—and the kingdom possessed extensive deposits of the principal raw materials required for the manufacture of cement. In addition, there was clearly no difficulty for Siam in acquiring the technical expertise and plant that would ensure the production of cement of at least equal quality to that of imported brands.[30] But these conditions, in themselves, cannot explain the success of the Siam Cement Company in rapidly achieving a domination of the domestic cement market. The crucial consideration here was that imported cement had to bear high freight charges, high relative to its production costs; consequently, imported cement entered the Siamese market at a distinct price disadvantage. As the Siam Cement Company prepared to start production, its retail price was calculated at 5.50 baht a barrel, compared to the current import price of over 6.00 baht a barrel.[31] A comparable price advantage was maintained at least until the early 1930s when, for a brief period, Siam became the victim of Japanese 'dumping'.[32] In short, the Siam Cement Company possessed a crucial measure of non-tariff protection against foreign competition.

In contrast, the Siamese Steamship Company, the second major business initiative of the government élite in this period, possessed no comparable form of protection against foreign competition—and it soon failed.[33] The company was inaugurated in January 1918, its initial capital of one million baht being provided equally by the government on the one hand, and a small group of prominent Siamese and Sino-Siamese entrepreneurs (including Suvabhan Sanitwongse) on the other. It was formed partly to assist the development of Siamese trade and partly to provide an auxiliary force for Siam's navy in times of war. Its original fleet comprised three German merchant vessels, seized by the Siamese Government when Siam had declared war on the Central Powers in July 1917. During its first year of operation, the Siamese Steamship Company was able to take advantage of a buoyant export trade and of the severe shortage of foreign shipping in Eastern seas due to the European war, to secure a profit of over 400,000 baht, declared in March 1919. In that same year, four new vessels were acquired. But in 1920 its position weakened dramatically. In that year, a failure of the rice crop caused the government to prohibit the export of rice; foreign shipping had returned to Bangkok; and the company now had four new vessels to be kept in operation. Hostile trading conditions were compounded by disaster at sea. The Siamese Steamship Company's largest vessel had foundered on rocks off the China coast in February 1918, and in December 1920 another vessel, its cargo uninsured, was lost off the Siamese coast. In that last year the company declared a deficit of 874,000 baht. In late 1920 the company undertook a major share issue in an attempt to rescue a rapidly deteriorating financial position; a further issue took place the following year.[34] Then, in January 1922, on the initiative of the Minister of Finance, Prince Chanthaburī, the British-owned Borneo Company was appointed managing agent of the Siamese Steamship Company. But the new management could do little to solve the fundamental commercial weaknesses of the concern. It continued to lose money. In 1926 it was wound up.

Part of the reason for the failure of the Siamese Steamship Company appears to have been the inadequate condition and structure of its fleet. Its original ships were aged and in a poor state of repair, for they had been laid up in Bangkok from August 1914 until they had been seized by the Siamese authorities three years later. Moreover, none of the company's ships was specifically designed for a particular trade, and only two were in any way equipped for the lucrative coolie trade between Bangkok and the South China ports. However, it can be argued that even with a modern, appropriately structured fleet, the Siamese Steamship Company would still have experienced considerable difficulty in competing on the centrally important runs from Bangkok to Singapore and from Bangkok to the South China ports, notably Hong Kong. It should first be noted that within two years of its establishment, the company faced a world-wide collapse of

THE PROMOTION OF ECONOMIC DIVERSIFICATION   157

shipping rates that put considerable financial pressure on even the major, long-established Western companies.[35] But more important, the structure of shipping routes in the region in that period appears to have been prejudicial to the profitable operation of a small, local line such as the Siamese Steamship Company. The latter could easily secure cargoes of rice in Bangkok for shipment to Singapore and Hong Kong, but it experienced very considerable difficulty in obtaining inward cargoes for Siam.[36] The reason for this was almost certainly that the major Western shipping companies which brought cargoes of manufactures from Europe into, in particular, Singapore, employed their associate or subsidiary lines when those cargoes were transhipped for outlying ports in the region such as Bangkok.[37] Denied inward cargoes in this way, the Siamese Steamship Company could not remain profitable on its operations overall.

II

In contrast to the establishment of domestic cement production and the abortive attempt to create a local shipping company, the two further forms of economic diversification to be considered in the remainder of this chapter, the modernization of domestic sericulture and the expansion of local cotton cultivation, involved the Siamese government élite not as entrepreneurs and individual investors but simply as administrators.

At the beginning of the twentieth century, virtually all the silk produced in Siam came from the north-east region of the kingdom. It is not possible to calculate with any confidence Siam's total production for this period, simply because a major (but indeterminate) proportion of that production did not enter the market but was consumed by the producing households themselves. However, in the early 1900s, it was estimated that in *monthon* Khōrāt some 20,000 households (between 40,000 and 60,000 people) were engaged in rearing silkworms, total annual production being in the order of 200 piculs.[38] Another contemporary source reported that there were in the town of Khōrāt 23 merchants dealing in raw silk, and estimated that each year some 2,000 piculs of raw silk were brought into the town from the surrounding provinces.[39] But there is also evidence from this period that silk production was contracting. At the end of the 1880s, a British consular report observed that 'the rearing of silk, beyond what is required for local wants, appears to be an occupation of diminishing importance'.[40] And during the final two decades of the nineteenth century, there was a very substantial increase in silk imports, principally of silk piece goods.[41] In brief, it would appear that at the opening of the twentieth century, Siam possessed a modest sericulture industry, but one which had been rapidly losing ground to imports.

In late 1900 two Japanese agriculture officials undertook an inspection tour of the sericulture districts of *monthon* Khōrāt, the tour being undertaken on the initiative of the Japanese Minister in Bangkok.[42] A copy of their report, drawing attention to the backward condition of the local industry and offering some preliminary suggestions for reform, was sent to Prince Damrong, and it was at the prompting of the Minister of the Interior that in early 1901 the Siamese Government decided to engage a Japanese sericulture authority to conduct a more detailed investigation. That authority, Kametaro Toyama, assumed his appointment in March 1902.[43] His initial report, submitted to the Minister of Agriculture the following month, starkly detailed the deficiencies of silk production in Siam: no care was taken in the cultivation of mulberry trees, with the result that the leaves were notably inferior; the silkworms were very small, and consequently produced very small cocoons; almost all silkworms were seriously diseased; there was no separation out of diseased and perforated cocoons prior to reeling; reeling implements were primitive.[44] But Toyama also confidently declared that the quantity and quality of silk produced in Siam could be markedly improved, drawing attention to the important advantages the kingdom possessed in having a diligent female work-force and a climate well suited to the culture of silkworms and the growing of mulberry trees. Thus encouraged, in early 1903, a separate Sericulture Department was created within the Ministry of Agriculture.[45] The Department's director was Prince Phenphat; Kametaro Toyama was retained as principal adviser.

The principal tasks of the new Department in its first years were to undertake series of detailed sericulture trials and experiments and to train Siamese sericulture instructors. The former required the establishment of a laboratory and mulberry plantation in the Prathumwan district of the capital.[46] To judge by the large number of weighty technical reports produced from the laboratory in this initial period, this experimental work drew a very strong commitment from Kametaro Toyama and his Japanese assistants. By late 1905, they had successfully cross-bred a superior strain of silkworm from Siamese and Japanese varieties, which was then to be distributed in the sericulture districts of the north-east.[47] The instruction of Siamese apprentices in basic sericulture techniques began in July 1903, skilled reelers and weavers being engaged from Japan as demonstrators.[48] In January 1905, a Sericulture School was opened, its principal concern being to train Siamese sericulture instructors who would on graduation replace their Japanese teachers.[49] This work did not proceed without its share of difficulties. In particular, teaching was initially hindered by the lack of textbooks in Siamese, although within a year the Department itself had begun to compile basic texts in the vernacular.[50]

The extension of the Sericulture Department's work to the provinces began with the establishment of a branch in Khōrāt town in 1904.[51] The following year a second branch was established in Burīram district, some 60 miles to the east. Both branches contained a mulberry plantation, a shed for rearing silkworms, and facilities for reeling silk; at a later point the Khōrāt branch developed facilities for silk-weaving. It was intended that each year a limited number of silk producers from the surrounding districts would attend a one-year course at the Khōrāt or the Burīram branch.[52] On completion of the course they would return to their home villages where, by direct instruction and by example, they would disseminate the more advanced sericulture practices they had learnt. But this initiative found a poor response. Few volunteered for the instruction offered—and among those who did come forward and complete the course, there appears to have been an unwillingness to impart their newly-acquired skills to others in their community on their return.[53] Consequently, from late 1907 the Sericulture Department began to take its instruction directly into the sericulture districts themselves, in time establishing eight sub-branches at principal locations throughout the north-east. Trained female silk-reelers were assigned to each sub-branch, with the task of inducing or cajoling the local women into taking up their more advanced practices. Simple reeling equipment was made available to all those learning the new methods. Finally, male graduates of the Sericulture School in Bangkok undertook tours of the north-east, providing instruction to all in the cultivation of mulberry and in the rearing of silkworms. This concentration of resources on the silk districts themselves towards the end of the 1900s was accompanied by a substantial reduction in the Bangkok establishment. In 1908 the original Bangkok Department was broken up, its administrative, research and teaching divisions being separately transferred to other departments in the capital or to the Khōrāt sericulture branch.

However, just as the administration's sericulture initiative came to be concentrated in the producing districts, it appears to have lost momentum. A crucial influence here was undoubtedly the death of Prince Phenphat in November 1909,[54] for as director he had provided a notably innovative and firm leadership from the programme's inception. It was also in the late 1900s that the number of Japanese sericulturists engaged on the programme was sharply reduced, their places being taken by Siamese sericulture graduates. At the end of 1908 only three Japanese remained; the last left in mid-1912.[55] By this time the sericulture programme had been greatly reduced; in early 1913 it was abandoned altogether.[56] Some 20 years later, a European official in the Ministry of Commerce and Communications reported that silk production in the north-east remained small-scale, technologically backward, and only marginally directed towards the market.[57] There was little to show for the government's initiative at the start of the century.

Why did the sericulture programme fail? The senior Siamese officials closely involved in the initiative focused on three broad considerations. It was noted that at first silk producers in the north-east had been suspicious of this major intervention by central authority.[58] Important here was the fact that the early Japanese instructors, with a poor command of the Siamese language, were unable to communicate clearly the administration's intentions, for this had allowed the belief to take hold in the silk districts that the government was in fact intent on introducing a new form of corvée obligation. In addition, there was much distrust of the Japanese—simply as foreigners—it being said that they had come to the north-east to dupe the local people. On the other hand, this wariness was soon dispelled as local Siamese officials were brought in to explain clearly to the leaders of each community the aims of the sericulture programme, as the superiority of the new practices was widely demonstrated, and as Siamese sericulture graduates replaced their Japanese teachers in the provinces. Second, and more important, senior Siamese officials frequently complained that the peasants in the north-east simply did not possess the qualities of character to respond to the administration's initiative. At various times the latter were seen as credulous and timid, as having limited material ambitions, or as being indolent.[59] A comparably disparaging view was later advanced by W. A. Graham, when he argued that the 'earnest endeavours' of the administration had failed in the face of the apathy, indifference, and 'astonishing lethargy' of the people.[60]

Hence the young women who had undergone training at the Government school, and had incidentally received a subsistence allowance from the State while doing so, on returning to their homes divested themselves as soon as possible of any knowledge they had acquired and, if they went in for silk at all, adopted the ways advocated by their grandmothers; while the new-fangled foreign implements given them on leaving school were stuck up in the thatch of the paternal cottage, where it was hoped that any foreign magic adhering to them might bring general good luck to the family.[61]

The final consideration was one strongly emphasized by Prince Phenphat and by Chulalongkorn himself: it was the view that all too frequently the administration's sericulture initiative had been crucially undermined by the overbearing, haughty manner of the provincial sericulture officials. Writing at the close of 1908, the King suggested that

[Sericulture] instructors working in the rural districts have sought to coerce the local people. They provide no guidance for the people as to how they might make a living [from sericulture]. When a small misunderstanding arises they become angry and create a scene, as if they were noblemen or senior officials. It does them no good at all. On the contrary, the people come to detest them, become fed up with them. As a result, although the people might study [the new sericulture methods], they have been so antagonized that they stubbornly refuse to put them into practice. The success of sericulture instruction depends upon acquiescence and benevolence, upon seeking to explain [to the local people] the advantages [of the new methods]. It does not require coercion.[62]

But a major weakness of each of these views is that they see the failure of the sericulture programme of the early twentieth century essentially in terms of an inadequacy or perversion of human attitude and behaviour—officials were autocratic, peasants simply irrational. In doing so, they ignore a number of basic technical and economic considerations.

Two observations are important here. First, the government's initiative sought only the most modest advance in sericulture practices in the kingdom. Of central importance here was the administration's approach to the reform of reeling methods. In January 1903, as the Sericulture Department was being established,

Kametaro Toyama had strongly argued that if local silk was to find a substantial market, it was essential to produce in large quantities a uniform, high-grade, yarn, and that this could be achieved only by employing steam-powered machine reelers.[63] In terms of both the quantity and uniformity of yarn produced, hand reelers were clearly inadequate in these circumstances—they should be confined to reeling for consumption within the producing household. Kametaro Toyama thus proposed that the government establish a model machine-reeling factory to train Siamese in those more advanced practices: the factory would also act to encourage an increase in cocoon production by the peasants in the north-east. That proposal was initially accepted; but it was later abandoned, as the Sericulture Department decided to concentrate its effort here on the advance of hand-reeling practices within the peasant household.[64] In broad terms, the Department's work in the north-east sericulture provinces from the mid-1900s was built on the premise that the existing structure of the industry, whereby each household undertook all stages of production from the cultivation of mulberry leaves to the weaving of silk fabrics, would be retained: it did not envisage the emergence of a functionally disintegrated industry, with certain stages (notably reeling) being concentrated in specialist, technically-advanced establishments.

Thus Siam's silk industry was not to pursue the structural changes then being effected in other Asian silk producers. For example, in Japan the first decade of the twentieth century saw a sharp rise in the proportion of silk yarn reeled by machine to 72 per cent of total output in 1909–13, and a concomitant emergence of large-scale reeling establishments.[65] A distinct, substantial, machine-reeling sector also emerged in China from the late nineteenth century.[66] In Indo-China, a silk-spinning and weaving factory was established by French capital at the close of the nineteenth century in Binh Dinh province, central Annam, although it might be noted that in time the weaving section of that concern, and of a second factory later established at Nam Dinh in the Tonkin delta, secured its yarn from mills at Shanghai and Canton.[67] A further feature of the Indo-China industry, important in this context, is that it was common for a village in the sericulture districts to specialize in just one aspect of silk production.[68]

The Siamese administrative records do not adequately explain why the sericulture authorities did not encourage the functional disintegration of the local industry. In reporting the Sericulture

Department's decision to abandon the proposal to establish a model machine-reeling factory, Prince Phenphat referred to (unspecified) difficulties that would arise in training Siamese on machine reelers; the inappropriateness of that training, in that it could not be put to use in the peasant's home; the difficulties that would be encountered in gathering in sufficient cocoons to satisfy the large capacity of a machine-reeling establishment; and the large investment that would be required of the government.[69] In broader terms, the administration may not have wished to see in the outlying districts of the north-east the disruption of economic and social structures that almost inevitably would have accompanied a major reorganization of silk production. Alternatively, it may not have fully grasped that the creation of a functionally disintegrated industry was simply essential if local silk was to find a substantial market; or again, the administration may not in fact have been particularly concerned with the market potential for local silk, but wished in effect to advance non-market production, that production which was consumed within the producing households. But the consequence of the administration's basic approach in the reform of the industry (whatever its precise perceptions and ambitions) was quite simply to leave the silk producers of the north-east with little prospect of being able to compete against producers in other parts of Asia. It was not only that the major advance of machine reelers elsewhere had secured a large-scale production of a uniform, high grade, yarn (as Kametaro Toyama had argued in 1903); functional disintegration had permitted the development of standardized practices at all stages of silk production. As a result, during the decade in which the Siamese Government had sought to reform (too modestly) the north-east silk industry, the value of raw silk and silk piece-goods imports into Siam, principally from China, Japan, and Indo-China, virtually doubled.[70]

The second consideration must be dealt with more briefly. It is the argument that the responsiveness of the north-eastern silk producers to the government's initiative was also influenced by the relative attractiveness of alternative forms of market production. Unfortunately, there does not exist the detailed data on agricultural costs and commodity prices in the north-east in this period to make possible an adequate exploration of this point. However, it can be noted that from the close of the nineteenth century the north-east, with its abundant grassland, emerged as the principal region in

Siam for the raising of water buffalo and cattle, a substantial inland trade developing with the Central Plain and an export trade being established to Singapore.[71] In other words, the region's comparative advantage at this time clearly lay in the production of livestock, and as market opportunities arose the villagers there enthusiastically exploited it. It need be added only that there was a notable rise in cattle prices, and in exports, just at the time the administration's sericulture programme was being extended into the north-eastern provinces.[72]

At the same time as the sericulture programme was being dismantled, the administration embarked on a further, much smaller, initiative to promote the cultivation of cotton in *monthon* Phitsanulōk, in the northern part of the Central Plain.[73] Prior to the opening of the kingdom to unrestricted foreign trade in the middle of the nineteenth century, the cultivation of cotton, as well as domestic spinning and weaving, had flourished in that area: but then as inexpensive, uniform cotton cloth and yarns flooded into Siam from the industrialized West and as local peasant households were drawn into an acute specialization in the cultivation of rice, cotton cultivation in Phitsanulōk collapsed. By the beginning of the twentieth century the acreage there under cotton was negligible. Seeking to revive cultivation, from 1912 the Ministry of Agriculture began to distribute higher grade cotton seeds acquired from Cambodia to the Phitsanulōk farmers and to despatch experienced agriculture officials to the *monthon* to offer advice and assistance in the cultivation of the crop. As production began to expand, in early 1913 local merchants combined to establish a cotton-ginning plant in Phičhit district, the seeds which the plant separated out then being distributed to farmers, thus facilitating further expansion of acreage. Later the same year the Ministry of Agriculture established its own ginning plant in Phitsanulōk. By 1914 there were 9,000 *rai* in the *monthon* planted with cotton, compared with 1,800 *rai* four years earlier.[74] At harvest, increasing numbers of merchants were coming into the area to purchase the crop, and the provincial administration began to see cotton as a potentially important export.

But these ambitions were destroyed by the outbreak of war in Europe in August 1914 and the consequent collapse of the world market in raw cotton. During the following harvest season, not a single merchant came into Phitsanulōk to purchase the *monthon*'s crop. The local cultivators were severely distressed, for with very few exceptions none had the resources or facilities to store their output adequately until the merchants returned. They were left holding a rapidly deteriorating crop. The collapse of demand also led to the closure of the cotton-ginning plant in Phičhit district, and a sharp reduction in the level of activity at the government's establishment. This, in turn, further threatened the future of cotton cultivation in the *monthon* by tightly contracting the production of seed.

The Superintendent Commissioner of *monthon* Phitsanulōk immediately urged government intervention.[75] He proposed that the administration rapidly buy up and gin the entire cotton crop of the area. This would dissuade the local farmers from abandoning the cultivation of cotton, as well as secure the future supply of seed. He further proposed that the administration construct and equip a modern spinning plant. This would ensure that the income from the spinning of Siam's raw cotton would be retained within the kingdom rather than, as was the current position, being allowed to fall into the hands of spinning companies overseas; more importantly, it would free local farmers from the vagaries of the world raw cotton market by providing a stable internal demand, and thus encourage a firm expansion in the cultivation of the crop. These proposals were firmly supported by the Minister of Finance, Prince Čhanthaburī.[76] However, there was strong opposition from Prince Rātburī, the Minister of Agriculture, on the grounds that the government would almost certainly suffer a heavy financial loss were it to assume responsibility for that year's cotton crop; that the proposed measures would establish a precedent for the producers of other crops to call for government assistance should they in turn experience serious distress; and, most interestingly, that cotton could never become a major crop in Siam, like, for example, rice, and therefore that the government should not encourage its cultivation.[77] Indeed the Minister felt it necessary to admit, and with some embarrassment, that in urging farmers to take up the crop earlier in the decade, his Ministry had been seriously in error. With this declaration, the prospects for government intervention disappeared. Thus denied a secure domestic demand, and already dispirited by the collapse of their external market, the kingdom's farmers declined to expand their commitment to cotton beyond its existing level—that is, until the demands of the Greater East Asia Co-Prosperity Sphere towards the close of the 1930s provided the secure market they sought.[78]

III

A common argument has underlain this chapter's consideration of economic diversification in early twentieth-century Siam: it should now be made explicit for it will be a crucial element in the following, concluding chapter. For each of the forms of economic diversification considered here, the founding of the Siam Cement Company and of the Siamese Steamship Company and the sericulture and cotton initiatives, the principal influence which determined success or failure lay not within the internal economic structures of the kingdom but in the structures of the external economy. Thus the cotton initiative failed in the mid-1910s in the absence, as perceived by the cultivators, of a secure external demand; the sericulture initiative failed in the face of the more advanced structural organization and production methods of silk industries elsewhere in Asia; and the shipping venture failed in the face of the highly-integrated shipping networks created by the major Western lines operating in Eastern seas; while the cement venture succeeded principally because locally-produced cement was assured a price advantage over imports in the domestic market. In other words, the extent and direction of the economic diversification achieved in Siam in the early twentieth century was determined principally by the powerful and diverse influences of the world economy: it was an inevitable reflection of Siam's open integration into the world economic community from the middle of the nineteenth century.

1. Board of Directors of the Siam Portland Cement Company to Chulalongkorn, 6 July 1909, NA r5 KS 12.2/16.
2. HM Customs Department, Bangkok, *The Foreign Trade and Navigation of the Port of Bangkok* (1910–11 and 1911–12).
3. E. G. Gollo, 'Report on Siam Portland Cement', 7 June 1909, NA r5 KS 12.2/16.
4. Čhaophrayā Yomarāt, 'Lap chaphǫ čhotmāibanthu'k ru'ang tangbǫrisatcement' ('Confidential Notes on the Establishment of the Cement Company'), 20 August 1913, NA r6 RL 20/13.
5. The Siam Portland Cement Company Ltd. Prospectus. [July 1909]. NA r5 KS 12.2/16.
6. Ibid.
7. Board of Directors of the Siam Portland Cement Company to Chulalongkorn, 6 July 1909, NA r5 KS 12.2/16.

8. Chulalongkorn to Prince Sommot, 8 July 1909, NA r5 KS 12.2/16.
9. E. G. Gollo, 'Report on Siam Portland Cement', 7 June 1909, NA r5 KS 12.2/16.
10. Chulalongkorn to Prince Sommot, 8 July 1909, NA r5 KS 12.2/16.
11. Virginia Thompson, *Thailand: The New Siam*, New York, 1941, p. 451.
12. Čhaophrayā Yomarāt, 'Lap chaphǫ čhotmāibanthu'k', 20 August 1913, NA r6 RL 20/13.
13. Ibid.
14. Čhaophrayā Yomarāt to Vajiravudh, 4 April 1914, NA r6 RL 20/13. It was understood that after 5 years, ownership of those shares would be transferred to the Privy Purse. But in April 1919 the latter accepted Čhaophrayā Yomarāt's Sālādaeng house as alternative security; at the same time the Minister undertook to repay the Privy Purse loan over 6 years (Čhaophrayā Yomarāt to Vajiravudh, 8 April 1919, NA r6 RL 20/13).
15. Čhaophrayā Yomarāt, 'Lap chaphǫ čhotmāibanthu'k', 20 August 1913, NA r6 RL 20/13. Grut subsequently persuaded a number of his associates to take up a portion of these shares (Čhaophrayā Yomarāt to Vajiravudh, 4 April 1914, NA r6 RL 20/13).
16. Čhaophrayā Yomarāt to Vajiravudh, May 1913, NA r6 RL 20/13.
17. *The Siam Cement Company: In Commemoration of the 50th Anniversary*, Bangkok, 1963, pp. 125-6.
18. Anders Tandrup, 'Some Danish Contributions to the Administrative and Socio-economic Development of Thailand since 1875', in *Thai-Danish Relations: 30 Cycles of Friendship*, Copenhagen, 1980, p. 172, note 44.
19. Oscar Schultz to Čhaophrayā Yomarāt, 7 May 1915, NA r6 RL 20/13.
20. *The Siam Cement Company*, Bangkok, 1957, p. 12.
21. HM Customs Department, Bangkok, *The Foreign Trade and Navigation of the Port of Bangkok* (various issues); HM Customs Department, Bangkok, *Annual Statement of the Foreign Trade and Navigation of the Kingdom of Siam* (various issues).
22. Ibid.
23. 'Industrial Siam: The Siam Cement Company', *The Record*, no. 9, July 1923, pp. 13-15; The Siam Cement Company, 'Balance Sheet Presented to the Annual Ordinary General Meeting of Shareholders', for the years 1922-4, 1929.
24. *The Siam Cement Company*, Bangkok, 1957, p. 9.
25. Ibid., p. 34.
26. See, for example, Vajiravudh to the Director of the Privy Purse Department, 29 June 1921, NA r6 RL 20/13.
27. The Siam Cement Company, 'Balance Sheet Presented to the Annual Ordinary General Meeting of Shareholders' (various years).
28. Čhaophrayā Yomarāt to Vajiravudh, 8 April 1919, NA r6 RL 20/13.
29. Ibid.
30. Experimental tests undertaken in London in 1917 apparently indicated that the local cement was distinctly superior to the brands then being imported into the kingdom. (Board of Directors of the Siam Cement Company to Vajiravudh, 6 July 1917, NA r6 RL 20/13.)
31. E. G. Gollo to Čhaophrayā Yomarāt, 26 March 1914, NA r6 RL 20/13.
32. Documents in NA K Kh 0301.1.7/31.
33. Unless otherwise noted, this section draws on, Stephen L. W. Greene, 'Thai

Government and Administration in the Reign of Rama VI (1910–1925)', Ph.D. diss., University of London, 1971, pp. 334–42; Walter F. Vella, *Chaiyo! King Vajiravudh and the Development of Thai Nationalism*, Honolulu, 1978, p. 172; W. J. F. Williamson, 'Note by the Financial Adviser on the present position of the Siamese Shipping Company; and an opinion as to the policy to be adopted towards it by the Government', 3 November 1922, NA r6 RL 20/22.

34. By 1922 the founding Siamese and Sino-Siamese entrepreneurs had substantially reduced their interest in the company; the principal shareholder was the Royal Navy League of Siam (an officially-sponsored body, formed in late 1914 to organize a public subscription of funds to purchase a light cruiser for the Navy); the Ministry of Finance held the government's original interest; and both the Privy Purse and the King's favourite were important shareholders.

35. S. G. Sturmey, *British Shipping and World Competition*, London, 1962, pp. 56, 65.

36. W. J. F. Williamson, 'Note by the Financial Adviser on the present position of the Siamese Shipping Company; and an opinion as to the policy to be adopted towards it by the Government', 3 November 1922, NA r6 RL 20/22.

37. Francis E. Hyde, *Far Eastern Trade 1860–1914*, London, 1973, pp. 92–3.

38. Kametaro Toyama to Chaophrayā Thēwēt, 8 January 1903, NA r5 KS 8/1 (one picul = 133⅓ lb).

39. *Bangkok Times*, 9 September 1902.

40. *DCR*, no. 771, 1890 [for 1889], p. 13.

41. *DCR* (annual issues for 1880s and 1890s).

42. Prince Damrong to Chulalongkorn, 12 February 1901, NA KS(Ag) 13/13.

43. Čhaophrayā Thēwēt to Chulalongkorn, 10 March 1902, NA r5 KS 8/1.

44. Kametaro Toyama to Čhaophrayā Thēwēt, 9 April 1902, NA r5 KS 8/1.

45. Čhaophrayā Wongsānupraphat, *Prawat krasuangkasētrāthikān* (*History of the Ministry of Agriculture*), Bangkok, 1941, p. 269.

46. Ibid., pp. 266–7.

47. 'Rāingān kromchāngmai' ('Report of the Sericulture Department'), August–November 1905, NA r5 KS 8/2.

48. Kametaro Toyama to Phrayā Srīsunthōnwōhān, 23 September 1903, NA r5 KS 8/1.

49. 'Rāingān kromchāngmai' ('Report of the Sericulture Department'), rs 123 [1904–5], NA r5 KS 8/2.

50. 'Rāingān kromchāngmai' ('Report of the Sericulture Department'), April–July 1904, NA r5 KS 8/2.

51. Čhaophrayā Wongsānupraphat, op. cit., pp. 269, 272, 281.

52. 'Rāingān kromchāngmai' ('Report of the Sericulture Department'), rs 123 [1904–5], NA r5 KS 8/2.

53. Čhaophrayā Wongsānupraphat, op. cit., pp. 272–82.

54. *Bangkok Times*, 12 November 1909.

55. Ibid., 27 November 1908; Documents in NA r5 B 9/48.

56. *Bangkok Times*, 16 January 1913.

57. Reginald le May, *The Economic Conditions of North-Eastern Siam*, Bangkok, 1932.

58. Prince Phenphat to Chulalongkorn, February 1909, NA r5 KS 8/2.

59. Deputy Commissioner, *monthon* Khōrāt to Sericulture Officer, Burīram district, 30 June 1908, NA r5 KS 8/2; 'Rāingān kromchāngmai' ('Report of the

THE PROMOTION OF ECONOMIC DIVERSIFICATION 169

Sericulture Department'), August–November 1905, NA r5 KS 8/2.
60. W. A. Graham, *Siam*, London, 1924, vol. 2, pp. 88–9.
61. Ibid.
62. Chulalongkorn to Prince Phenphat, 27 December 1908, NA r5 KS 8/2.
63. Kametaro Toyama to Čhaophrayā Thēwēt, 8 January 1903, NA r5 KS 8/1.
64. 'Rāingān kromchāngmai' ('Report of the Sericulture Department'), rs 123 [1904–5], NA r5 KS 8/2.
65. G. C. Allen, *A Short Economic History of Modern Japan*, 3rd ed., London, 1972, pp. 65–8.
66. Ramon H. Myers, *The Chinese Economy: Past and Present*, Belmont, California, 1980, p. 132.
67. Charles Robequain, *The Economic Development of French Indo-China*, London, 1944, pp. 282, 294.
68. Ibid., p. 246.
69. 'Rāingān kromchāngmai' ('Report of the Sericulture Department'), rs 123 [1904–5], NA r5 KS 8/2.
70. *DCR* (annual issues for 1900s and early 1910s).
71. Wolf Donner, *The Five Faces of Thailand: An Economic Geography*, London, 1978, pp. 148, 150, 617.
72. *DCR* (annual issues for later 1900s and early 1910s).
73. Except where indicated, this and the following paragraph draw on, The Superintendent Commissioner, *monthon* Phitsanulōk, 'Rāingān kānbamrungfāi' ('Report on the promotion of cotton'), 18 April 1915, NA r6 Kh 1/32.
74. Prince Čhanthaburī to Vajiravudh, 26 May 1915, NA r6 Kh 1/32.
75. The Superintendent Commissioner, *monthon* Phitsanulōk, 'Rāingān kānbamrungfāi' ('Report on the promotion of cotton'), 18 April 1915, NA r6 Kh 1/32.
76. Prince Čhanthaburī to Vajiravudh, 26 May 1915, NA r6 Kh 1/32.
77. Prince Čhanthaburī to Vajiravudh, 6 August 1915, NA r6 Kh 1/32.
78. James C. Ingram, *Economic Change in Thailand 1850–1970*, Stanford, 1971, p. 51; Thompson, op. cit., pp. 395–7.

Conclusion:
The Élite and the Economy in Siam,
c.1890–1920

It is now possible to return to the central questions which were raised in the introduction. To what extent did late nineteenth- and early twentieth-century Siam, under the authority of an indigenous administration, have an opportunity to pursue a pattern of economic change markedly different from the one which did emerge in the kingdom in this period (and in Siam's neighbours under European rule)? Did the Siamese administrative élite perceive either the necessity for, or the desirability of, promoting a markedly different pattern of economic change? To what extent and in what way was the administration constrained by either external or internal circumstances (including the self interest of its own governing élite) from promoting an alternative pattern of economic change? Had the Siamese administrative élite of this period, after five decades of increasing familiarity with the Western world, assumed the values, attitudes and intellectual horizons of the West to the extent that, in the promotion of economic change, it pursued essentially the same measures and sought substantially the same objectives as a European colonial administration? Alternatively, was the administrative élite in any way critical of the pattern of economic change which had emerged in the kingdom; and if so, to what extent and in what manner was it constrained from promoting an alternative pattern?

I

These questions have also been of central concern to the vigorous community of Thai historians, among whom Chatthip Nartsupha is the most influential, which has come to prominence from the early 1970s. It is not possible in this limited space to convey the full range of their argument or the way in which it has evolved and is evolving; the following paragraphs seek only to establish the principal elements in their response.[1]

Central to it is the assertion that in the nineteenth and early twentieth centuries, the king (frequently working through a wider royal and noble élite) controlled the major part of Siam's land, labour, and capital resources. In detail, most of the fertile land in the rice plain of Central Siam was owned by the Crown, the royal family, and noblemen, the last two acquiring their holdings through the generosity of the king. The Crown exercised a comparably formidable control over the kingdom's manpower, by requiring all males aged 18–60 to provide three months' labour in each year or, in some districts, contribute an appropriate payment in cash or in kind. And the Crown possessed the principal concentration of capital funds in Siam—it received a major flow of income from its holdings of land and its control of labour; and with Chulalongkorn's establishment of a central revenue office in the early 1870s, its control over the kingdom's tax revenues was consolidated.

This concentration of economic power in the hands of the king and the royal and noble élite (the *saktina* class) meant that that class absorbed a severely disproportionate share of the kingdom's surplus product, while the cultivators, the essential creators of wealth, were reduced to bare subsistence. For the latter, their impoverishment made it impossible for them to accumulate sufficient capital to raise the productivity of rice cultivation or to diversify into other forms of agriculture and into non-agricultural production. The suppression of occupational diversification was reinforced by the court's tight control over the work of indigenous craftsmen, for it prevented their evolution into an independent industrial bourgeoisie. For their part, the *saktina* élite used the vast resources at their command to secure their own dominant position and largely ignored the wider interests of the people as a whole. Thus in the late nineteenth and early twentieth centuries, government expenditure was heavily concentrated on defence, internal security, and the court: only relatively small allocations were made to education or to developmental projects, notably the construction of irrigation works.

Also of central importance in this analysis is the characterization of the kingdom's Chinese merchant community. It is argued that the flood of Western manufactures into Siam from the middle of the nineteenth century prevented those Chinese from establishing the industrial interests that would have provided them with a direct control over surplus production. As a result, they were forced to rely on the *saktina* élite for access to surplus production, and in this way were drawn into a subordinate relationship with it. It was a dependent relationship strongly sought by the Siamese élite itself: Chinese were practically exempt from labour obligations and were more lightly taxed than the indigenous population; unlike the latter, they were free to travel and trade throughout the kingdom; and prominent Chinese were appointed by the Crown to the kingdom's principal tax-farms, positions which carried with them a considerable measure of administrative and judicial authority, as well as an official rank which secured the holder within the Siamese hierarchy. In essence, the relationship centred upon the *saktina* élite's exchange of administrative protection and facility for an income of rents, interest and bribes; it was reinforced by inter-marriage and, ultimately, by the cultural assimilation of the Chinese. But, crucially, by being drawn into a subservient relationship with the *saktina* élite, the Chinese merchant community assumed the conservative, bureaucratic ideology of that class, faithfully serving its interests. They remained a dependent bourgeoisie, constrained from technical and organizational innovation and from the accumulation of capital.

It is now possible to consider in a little more detail the response of recent Thai scholarship to the two principal aspects of economic change in late nineteenth- and early twentieth-century Siam which have been of primary concern in this book—the failure of agriculture to secure a sustained increase in productivity or to achieve significant crop diversification; and the absence of major industrial growth. As noted above, the stagnation of productivity and the essential absence of diversification in agriculture is explained in part by the cultivators' impoverishment; they could not command the resources to invest in agricultural advance. It is further argued that advance was obstructed by the concentration of landownership in the hands of the *saktina* élite, for as absentee landlords they usually showed little interest in improving their holdings, while they discouraged their tenants from making improvements by demanding higher rent whenever the latter's initiative did secure an increase in yields. But a particular emphasis is placed here on the failure of the government. In the early twentieth century the Siamese administration may have discussed at length a wide range of proposals to improve agricultural practices in the kingdom, but no significant action was ever taken. The administration also repeatedly declined to undertake the construction of a large-scale irrigation facility in the lower Central Plain, although this would have effected a major increase in rice production and a substantial measure of agricultural diversification. When it was eventually forced by deteriorating agricultural conditions to commit funds to irrigation in the mid-1910s, the administration sanctioned a relatively modest project (in the Pāsak district) that secured the landholding interests of the *saktina* élite in Rangsit but contributed very little to increasing total rice production in the Central Plain.

With respect to Siam's failure to industrialize, Thai scholarship has dismissed as inadequate earlier interpretations which have emphasized the absence of tariff protection against imported manufactures, the small size of the domestic market, and shortages of raw materials, capital and entrepreneurship. Rather, as noted above, attention is drawn to the tightening domination of the Chinese merchant community by the *saktina* élite, a process which prevented the evolution of the former into an independent, innovative bourgeoisie that could effect a fundamental change in the structure of the Siamese economy. But, again, particular emphasis is placed on the failure of the government. In the early twentieth century, the government élite, with the partial exception of the Crown, is seen to have disposed of its acquired surplus largely in luxurious consumption, leaving little for capital investment in industrial initiatives; it is also seen to have possessed a conservative, bureaucratic mentality that encouraged only a minimum interest in trade and industry. Consequently, the government offered no assistance to any of the private initiatives and proposals (for a textile factory, a tannery, a tin-smelter) which were advanced in this period, and as a result those initiatives failed. When, in an exceptional case, the government did provide strong support—that is, to the Siam Cement Company—that initiative was successful: it was successful *because* of the government's strong support.

II

Thus, in seeking to understand the failure of Siam to achieve either agricultural advance or significant industrial expansion in the late nineteenth and early twentieth centuries, recent Thai scholarship has focused principally (but not exclusively) on the internal economic, social and political structures of the kingdom. In essence, the *saktina* élite's control of the kingdom's land, labour and capital resources denied the cultivators (the creators of wealth) the means to initiate fundamental economic change: yet that control, through the economic self-interest and restricted perceptions which it engendered, also left the élite itself incapable of pursuing fundamental reform, while causing it to stifle the evolution of the Chinese merchant class into an innovative bourgeoisie that could effect a decisive restructuring of Siam's economy. In contrast, the earlier chapters of this book have sought to emphasize the primary importance of external constraints in determining the nature of economic change in late nineteenth- and early twentieth-century Siam.[2] That discussion can now be drawn together.

For the Siamese administration, the potentially most important economic initiative in this period would have been the construction of major irrigation works in the lower Čhaophrayā plain, for this would have significantly raised the productive capacity of the overwhelmingly dominant rice economy, thus providing the resources for further agricultural advance and for industrial expansion. The argument advanced in recent Thai scholarship that in the early twentieth century the Siamese administration 'paid but a token attention to irrigation problems'[3] cannot be sustained in the face of the evidence presented earlier in this book. It ignores not only the length and intensity of the administration's discussion but, more importantly, the significant shifts which took place in this period in the government's perceptions and ambitions. At the turn of the century the enthusiasm of Chulalongkorn and Čhaophrayā Thēwēt for a major irrigation initiative may well have been inadequately considered and insecurely based. But by the early 1910s, Prince Rātburī had persuaded the administration into a secure appreciation of the arguments for the construction of major irrigation works in the Central Plain and of the precise benefits that initiative would yield; it was an appreciation that recognized the importance for the kingdom's continued prosperity of maintaining the competitiveness of Siamese rice in international markets. Neither is it possible to sustain the argument that the substantial landholding interests of the Siamese administrative élite in Rangsit blocked a major government irrigation initiative. The documentary evidence offered in support of the argument is uncomfortably slight, while the argument itself is open to objection on grounds of logic. The principal reason for the Siamese administration's failure to commit itself to the construction of major irrigation works in Central Siam in the early twentieth century was, to repeat an earlier conclusion, the high cost of that project in relation both to the volume of resources at the government's command and to the demands of the other major expenditure programmes before it. It is important to emphasize that these latter considerations—the volume of resources available to government and their allocation between competing expenditure demands—constituted constraints that were essentially externally imposed. The ability of the Siamese administration to raise tax revenue from foreign trade and from the cultivation of land (invariably the principal sources of government revenue in primary exporters in this period) was restricted by the treaties concluded between the kingdom and the major powers from the middle of the nineteenth century, while its willingness to raise capital on the finance markets of Europe was constrained by the fear (justified by the contemporary experience of China) that borrowing from the imperial powers could well bring with it a dangerous degree of foreign intrusion. And the allocation of those resources—towards the construction of railways, the strengthening of military forces, and the creation of substantial exchange reserves, but not towards irrigation—was determined by the overriding need to defend the political sovereignty of Siam, threatened by European territorial expansion and commercial aggression in mainland South-East Asia from the middle of the nineteenth century.

The failure of the Siamese administration to commit itself to a major investment in the improvement of agricultural inputs and methods (the development and diffusion of high-yield rice seeds, the introduction of more advanced agricultural equipment, the encouragement of an increased use of fertilizer, and the evolution of more productive cycles of rice planting and harvesting) might also be seen in the context of the alternative, inescapable demands on the government's restricted financial resources. At the same time, it is clear from the earlier discussion that for the administration to secure a significant improvement in agricultural inputs and practices posed a far more formidable challenge to its perceptivity and ingenuity than, for example, the construction of irrigation works. In essence, it required the creation of a body of distinctive agricultural knowledge specifically appropriate to Siamese conditions, a task which, involving as it did a tempering of the knowledge of Western scientific agriculture acquired by a younger generation of Siamese officials by the extensive experience of indigenous conditions and practices held by their seniors, demanded a social sensitivity between the generations apparently rarely found in the ranks of the administration at that time. But even then, evidence from elsewhere in Asia would suggest that the principal initiative for agricultural advance in this period did not after all lie with government but was found within the rural community itself. Thus, a relatively substantial investment in agricultural improvement undertaken by the colonial administrations in Burma and in the Philippines from the beginning of the twentieth century appears to have had a negligible impact on cultivation practices and inputs, whereas in Japan the primary initiative in agricultural innovation (at least in its crucial initial phase to the early twentieth century) was taken not by government but by the cultivators and landlords themselves. Here, exceptionally, the absence of change in Siam may well be explained primarily in terms of internal considerations—not government neglect but in the structure of social authority and economic power within the kingdom's rural communities which clearly lay beyond the immediate intervention of government.

Turning now to the two important extractive industries, teak and tin, the principal argument, again, is that a major restructuring lay beyond government intervention in this period. At the close of the nineteenth century, the primary concern of Bangkok here was to bring the exploitation of teak and tin by external (predominantly Western) capital under firm government regulation. This was a matter of political urgency: and it was successfully accomplished, essentially because it required only a comparatively modest administrative initiative. Yet government regulation did little to disturb the essential structures of those industries. Thus, the size and rate of growth of the peninsular tin industry, its ethnic structure, the pace of technological change within it, and its high and increasing dependence on British interests in Malaya for the smelting of its output were still predominantly influenced by the relative attractiveness of other tin-bearing areas for investment by Western capital, the changing technological demands of tin mining in Siam, the level of world demand for tin, and the economies secured through large-scale tin smelting. And with respect to teak, the advancing Western domination of the industry from the end of the nineteenth century arose from the commercial reality that an efficient large-scale working of the teak stands demanded an investment of fixed and working capital on a scale that at that time only the major European forestry companies appeared able to provide. With the northern forests and southern tin deposits open to Western investment and trade (and access was secured by the treaties concluded by the major powers with Siam from the middle of the nineteenth century), it lay beyond the power of the administration in Bangkok, commanding limited resources, to shape the basic structures of the teak and tin industries.

And finally, why did Siam fail to achieve a significant measure of industrial and commercial diversification in the late nineteenth and early twentieth centuries? Once again the argument advanced in recent Thai scholarship—here focusing on a powerfully restrictive government élite, bound within its bureaucratic, uncommercial mentality and intent on subordinating Chinese merchant capital to its established interests—cannot be sustained in the face of the evidence presented in the immediately preceding chapters. The founding of the Siam Commercial Bank, the Siam Cement Company, and the Siamese Steamship Company must surely be seen as evidence not only of firm commercial ambitions within the governing élite but also (at least in the case of the bank and the shipping line) of the élite's eagerness to encourage a commercial diversification of Chinese merchant capital in the furtherance of those ambitions. Rather than seeking to emasculate Chinese merchant capital, the Siamese government élite had a strong interest in its evolution, as long as it was held in alliance with *saktina* political authority. Thus, the Siam Commercial Bank and the Siam Cement Company were important new sources of wealth for members of the government élite: indeed, as was shown in the earlier discussion of Čhaophrayā Yomarāt's involvement in the cement company, it was only the income from such initiatives that enabled some ministers to maintain their social position at a time when the income from government service was coming under stricter regulation.[4] If, therefore, there was an essential convergence of private interest between Siamese political authority and Chinese entrepreneurial power in the creation of industrial and commercial enterprise in the kingdom, why did that process not proceed further than it did?[5] The argument developed here is that the extent and direction of industrial and commercial diversification was firmly limited by the openness of Siam to the powerful and diverse influences of the world economy—that a particular enterprise would succeed only to the extent that it could be protected from those influences. Thus the Siam Commercial Bank was successful primarily because the government could provide it with a large measure of protection against the established European banks in Bangkok. In detail, over the years the government placed a major part of its deposit and exchange transactions with the Siam Commercial Bank rather than with its European rivals, so providing the local bank with substantial secure business as well as signalling official confidence in it to the wider public. Moreover, in 1913 it was the government, alone, which prevented the collapse of the Siam Commercial Bank; and in the years which followed, the administration always stood behind the bank in times of crisis. The Siam Cement Company succeeded primarily because locally-produced cement was naturally assured a price advantage over imported cement in the Siamese market. In contrast, the Siamese Steamship Company possessed no comparable form of protection against foreign competition, and therefore it failed in the face of the highly-integrated shipping networks created by the major Western lines operating in Eastern seas. And the argument can be extended. The government did not pursue, for example, the proposal for the establishment of a tin-smelter at Phūket,[6] not because of its bureaucratic mentality or indifference to commercial initiative but simply because, with the world's major smelting capacity just a few hundred miles to the south and with Siamese tin production low (and to fall sharply in the early 1920s), it must have been starkly clear that a local smelter could not at that time survive. Any administration in Bangkok in this period, whatever its ideology and economic foundation, would have faced the same severe externally-imposed constraints on its ability to secure substantial industrial and commercial diversification.[7]

III

This argument may provoke the objection that from the middle of the nineteenth century, Japan too was opened by unequal treaties to the unrestrained influences of the world economy and yet, in sharp contrast to Siam, moved rapidly towards a fundamental restructuring of the domestic economy. Indeed that sharply divergent experience has recently drawn a number of scholars into detailed comparisons of nineteenth-century Japanese and Siamese political institutions, government administration, and social and economic structures, as well as broader comparisons of Thai and Japanese value systems.[8] But in fact the assumption upon which those comparisons proceed, that from the middle of the nineteenth century both Japan and Siam 'worked under basically the same external conditions',[9] is by no means firmly established; indeed it can be argued that the political and economic relationships of

Japan with the Western world in this period were fundamentally different from those of Siam. With respect to the political relationship, the contrast needs to be made between a Siam vulnerable to European armed intervention as colonial possessions pushed up to her borders and an insular Japan essentially secure from Western occupation. Siam's vulnerability, as noted earlier, had a dominating influence in the allocation of government expenditures from the later nineteenth century: but it also had a more pervasive and penetrating influence, shaping the fundamental perceptions and ambitions of the Siamese élite of that period, forcing it into an essentially deferential, submissive relationship to Western power. In contrast, Japan's security permitted, perhaps even encouraged her to contemplate equality with Western power, to drive towards the creation of a major military capability and a strong industrial base. With respect to the economic relationship, the essential argument is that in the second half of the nineteenth century it was far more difficult for Western manufactures to penetrate the Japanese than the Siamese market. For Japan, two related considerations were important here. Under the treaty port system, which remained in force until 1899, Western merchants were not permitted to travel beyond the designated ports in pursuit of trade; they could not, therefore, create their own distribution networks in the Japanese interior. At the same time, they found it very difficult to penetrate the tightly-organized, resilient Japanese distribution networks which had their origins in the Tokugawa period and which were further strengthened in the early Meiji years. This indigenous control of Japan's internal market provided a powerful non-tariff barrier against penetration by Western manufactures, securing an essential protection for emerging Japanese industry.[10] Thus, notably, as Western cotton manufactures swept into Siam, in effect destroying traditional cloth production in the Central Plain as well as the prospects for the creation of a modern textile industry, they made only slow progress in the Japanese market, thus permitting the rapid emergence of a modern cotton sector as a leading element in Japan's industrialization.[11]

This is not to deny that there were major differences between Japanese and Siamese political institutions, government administration, and social and economic structures in the nineteenth century as well as between Thai and Japanese value systems; nor that the distinctiveness of Japanese institutions, administration, structures and values provide an important part of the explanation for Japan's economic transformation in the second half of the nineteenth century. The argument here is that there was a fundamental difference in the manner and degree to which Siam and Japan were incorporated into the world economy from the mid-nineteenth century, and it is there, in the external relationships, that the prime explanation for their divergent economic experience is to be found.

IV

One issue remains. It is evident from earlier discussion that in the opening decades of the twentieth century, there were individuals within the Siamese administrative élite who were seriously disturbed by particular structural weaknesses in the kingdom's economy. Notable here were Čhaophrayā Wongsānupraphat, arguing in 1910 against Siam's heavy dependence on imports for a wide range of foodstuffs and household utensils which could well be produced domestically;[12] and Prince Phenphat and Mǫm čhao Sithiporn Kridakara, persistently concerned to drive the highly specialist rice communities of the Central Plain into more diversified cropping and livestock production.[13] Yet it is clear that these individual, particularistic concerns did not evolve into a fundamental questioning on the part of the administrative élite as a whole of the dominating influence of the international economy on the pattern of economic change in Siam.[14]

The failure of such a critical consensus to emerge may be explained in part by two related considerations. The Siamese administrative élite of the early twentieth century, when surveying the economic changes which had taken place in the kingdom over the preceding five decades, would perhaps have been more inclined to marvel at the transformation of the Central Plain into one of the world's most important rice-producing regions and the concomitant developments in international trade, transport facilities and commercial infrastructure, than have directed a critical, doubting eye at the structure which had emerged. As rice production and export grew at an unprecedented pace, as the world's manufactures became available in the most distant rural markets of the kingdom, as exchange banks, shipping offices and merchant houses flourished in Bangkok, was it not evident that Siam had been drawn into the modern world of economic progress? It is important to note here that from the opening of the kingdom to unrestricted international trade in the middle of the nineteenth century right to the disintegration of the world economy between the wars, the basic structure of Siam's economy, its long-term stability and advantage were not challenged by external depression. It is true that the kingdom experienced a number of short-term, moderately sharp recessions, as towards the end of the 1900s and again, more seriously, at the close of the following decade; but these contractions (at least that in the late 1900s)[15] were not of sufficient severity to provoke or force the Siamese administrative élite into a searching examination of the economy's fundamental structure or of the dominant determinants of economic change in Siam. In other words, although it was of course evident in this period that the kingdom had developed an acute dependence on the export of rice and had achieved only the most modest industrial expansion, the inability of that economic structure to secure long-term economic growth and increasing material welfare had not been established. This was to come into focus only with the world economic crisis of the 1930s and, perhaps more importantly, with the rising economic expectations which accompanied the European territorial retreat from Asia in the 1940s and 1950s.

It must be emphasized, moreover, that the pattern of economic change which emerged in Siam in the late nineteenth and early twentieth centuries simply reflected the economic orthodoxy of that time, an orthodoxy that held that economies in the non-European world would secure a firmly-based prosperity through strong specialization in the production and export of those (agricultural and mineral) commodities in which they had a comparative advantage. It was this economic orthodoxy that the Siamese administrative élite accepted, perhaps in part because it was also shaping the pattern of economic change in Siam's neighbours under European rule but also because, as argued above, from the middle of the nineteenth century it had indeed secured for the kingdom a quite unprecedented growth in economic activity. Yet within it lay a final external constraint, not here on the freedom of action of the Siamese administrative élite, but on their perceptions and horizons.

---

1. This section draws primarily on an early statement: Chatthip Nartsupha and Suthy Prasartset (eds.), *The Political Economy of Siam 1851–1910*, Bangkok, 1981; Chatthip Nartsupha, Suthy Prasartset, and Montri Chenvidyakarn (eds.), *The*

Political Economy of Siam 1910–1932, Bangkok, 1981; Suthy Prasartset and Chatthip Nartsupha, 'The Rise of Dependent Commodity Production in Siam, 1855–1910', *The Review of Thai Social Science*, 1977, pp. 144–68. From a very extensive literature in Thai, a major part comprising detailed empirical studies within the theoretical framework established by Chatthip and Suthy, particular note should be made of two compilations: Chatthip Nartsupha and Somphop Manarangsan (eds.), *Prawatsāt sēthakit thai čhonthu'ng p.s.* 2484 (*The Economic History of Thailand Before 1941*), Bangkok, 1984; *Ruambotkhatyō withayāniphon prawatsāt p.s.* 2488–2527 (*Abstracts of Theses in History, 1945–1984*), cremation volume for Professor Rong Syamananda, Bangkok, August 1985. There are two valuable English-language surveys of recent Thai historical scholarship: Craig J. Reynolds and Hong Lysa, 'Marxism in Thai Historical Studies', *Journal of Asian Studies*, vol. 43, no. 1 (November 1983), pp. 77–104; Hong Lysa, *Thailand in the Nineteenth Century: Evolution of the Economy and Society*, Singapore, 1984, chapters 6, 7.

2. It must be added that an emphasis on the primacy of external influences can also be found in some recent Thai scholarship (Reynolds and Hong, op. cit., pp. 92–3). Reference should also be made here to, Peter F. Bell, *The Historical Determinants of Underdevelopment in Thailand*, New Haven, February 1970.

3. Chatthip, Suthy, and Montri, op. cit., p. 13.

4. See p. 155.

5. The collaboration between Siamese political authority and Chinese capital in the creation of industrial and commercial enterprises (mainly for a later period) has been the subject of detailed study by Sungsidh Piriyarangsan, 'Thai Bureaucratic Capitalism, 1932–1960', Master of Economics diss., Thammasat University, 1980; *Thunniyom khunnāng thai p.s. 2475–2503* (*Thai Bureaucratic Capitalism, 1932–1960*), Bangkok, 1983. However, in seeking to assert that that collaboration 'obstructed the development of Thai capitalism' and, more particularly, to unravel the complex alliances which it involved, Sungsidh loses sight, I would argue, of the more important issue: why was that collaboration restricted to so few business initiatives; why did this form of commercial and industrial expansion not proceed further?

6. See pp. 108–9; Chatthip, Suthy, and Montri, op. cit., pp. 8–9.

7. This is an appropriate point to note that I have also developed this analysis in, 'The Siamese Administrative Élite in the Early Twentieth Century and the Historical Origins of Underdevelopment in Siam', in Jeremy H. C. S. Davidson (ed.), *Lāi Sū' Thai. Essays in Honour of E. H. S. Simmonds*, London, 1987, pp. 151–90.

8. For example, Likhit Dhiravegin, 'The Meiji Restoration (1868–1912) and the Chakkri Reformation (1868–1910): A Case for a Comparative Study'; Surichai Wun' Gaeo, 'The Significance of Socio-Political Factors in Meiji Economic Development with Comparative Reference to the Thai Case'; both in, Chira Hongladarom and Medhi Krongkaew (eds.), *Comparative Development: Japan and Thailand*, Bangkok, 1981, pp. 71–100 and pp. 101–29; Yasukichi Yasuba and Likhit Dhiravegin, 'Initial Conditions, Institutional Changes, Policy, and their Consequences: Siam and Japan, 1850–1914', in Kazushi Ohkawa and Gustav Ranis with Larry Meissner (eds.), *Japan and the Developing Countries: A Comparative Analysis*, Oxford, 1985, pp. 19–34; Eliezer B. Ayal, 'Value Systems and Economic Development in Japan and Thailand', in Robert O. Tilman (ed.), *Man, State, and Society in Contemporary Southeast Asia*, London, 1969, pp. 535–49. Reference might also be made to, Benedict R. O'G. Anderson, 'Studies of the Thai State: The State of Thai Studies', in Eliezer B. Ayal (ed.), *The Study of Thailand: Analyses of Knowledge, Approaches,*

and Prospects in Anthropology, Art History, Economics, History, and Political Science, Athens, Ohio, 1978, pp. 198–209.

9. Yasukichi Yasuba and Likhit Dhiravegin, op. cit., p. 20.

10. I am grateful to Shinya Sugiyama for this insight, drawn from his forthcoming *Japan's Industrialization in the World Economy, 1859–1899: Export Trade and Overseas Competition.*

11. Attention should also be drawn to Frances V. Moulder, *Japan, China and the Modern World Economy: Toward a Reinterpretation of East Asian Development ca. 1600 to ca. 1918*, Cambridge, 1977, pp. 128–45, to the argument that in the nineteenth century, Japan was incorporated into the world economy to a lesser degree than China, and that this contrast provides an important part of the explanation for their sharply divergent economic experience in the modern period.

12. Čhaophrayā Wongsānupraphat, 'Memorandum on our Domestic Economy', 7 December 1910, NA r6 KS 1/4.

13. Prince Phenphat to Čhaophrayā Thēwēt, 26 October 1904, NA r5 KS 10/1; Benjamin A. Batson, [Review Article], *M. C. Sithiporn Kridakara Memorial Volume, Journal of the Siam Society*, vol. 61, pt. 1 (January 1973), pp. 302–9.

14. Even Phrayā Suriyānuwat, whose *Sapsāt* (*The Science of Property*) written in 1911 constituted by far the most cogent critical analysis of the pattern of economic change in Siam in this period, did not pursue that crucial question, but focused instead on the internal maldistribution of economic power and income. (Chatthip Nartsupha, 'The Economic Thought of Phraya Suriyanuwat', in Vichitvong na Pombhejara (ed.), *Readings in Thailand's Political Economy*, Bangkok, 1978, pp. 402–13; Yuangrat (Pattanapongse) Wedel, *Modern Thai Radical Thought: The Siamization of Marxism and its Theoretical Problems*, Bangkok, 1982, pp. 50–8.)

15. See pp. 77–88.

Select Bibliography

Manuscript Sources

(a) **National Archives of Thailand (NA)**

(i) *Royal Secretariat (krom rātchalēkhāthikān)*
These records cover the period from the reform of government administration from the late 1880s to the end of the absolute monarchy. They comprise the administrative correspondence between the King (or his secretary) and the ministers, reports on meetings of the Council of Ministers, administrative notes and drafts made either by the King or his secretary, and copies of correspondence between and within ministries where these have been sent to the Royal Secretariat.
These records are arranged first by reign:

Chulalongkorn (	–1910):	r5
Vajiravudh	(1910–1925):	r6
Prajadhipok	(1925–1935):	r7

and then by ministry:

Ministry of Finance	Kh
Ministry of Agriculture	KS
Ministry of the Interior	M
Ministry of Public Instruction	S.Th

with two general classifications:

Miscellaneous	B and SB
Royal Secretariat	RL

and finally by subject, indicated by the numerical classification.
These records were by far the most important source for this study.

(ii) *Office of the Financial Adviser (samnakngān sēthakit kānkhlang)*
These are the records of the British Financial Advisers, the first of whom was appointed in 1896.
They carry the prefix: NA K Kh 0301.1 and are then divided by subject, indicated by the remaining numerical classification.

(iii) *Ministry of Agriculture (krasuang kasētrāthikān)*
These records of the Ministry carry the prefix: NA KS(Ag)

(b) **Public Record Office (London) (PRO)**

Foreign Office: General Correspondence, Political: FO 371

Newspaper

Bangkok Times (weekly, 1887–90; semi-weekly, 1891–4; thrice weekly, 1894–5; daily, 1896–1941)

Dissertations and Unpublished Papers

Battye, Noel Alfred, 'The Military, Government and Society in Siam, 1868–1910: Politics and Military Reform during the Reign of King Chulalongkorn', Ph.D. diss., Cornell University, 1974.

Brown, Ian, 'The Ministry of Finance and the Early Development of Modern Financial Administration in Siam, 1885–1910', Ph.D. diss., University of London, 1975.

Greene, Stephen L. W., 'Thai Government and Administration in the Reign of Rama VI (1910–1925)', Ph.D. diss., University of London, 1971.

Holm, David F., 'A History of the Teak Industry in Thailand', unpublished paper, Yale University, May 1969.

———, 'The Role of the State Railways in Thai History, 1892–1932', Ph.D. diss., Yale University, 1977.

Johnston, David B., 'Rural Society and the Rice Economy in Thailand, 1880–1930', Ph.D. diss., Yale University, 1975.

King, Frank H. H., 'The Foreign Exchange Banks in Siam, 1888–1918, and the National Bank Question', unpublished paper presented to the International Conference on Thai Studies, Bangkok, 1984.

Martin, James V., 'A History of the Diplomatic Relations between Siam and the United States of America, 1833–1929', Ph.D. diss., Fletcher School of Law and Administration, 1947.

Sungsidh Piriyarangsan, 'Thai Bureaucratic Capitalism, 1932–1960', Master of Economics diss., Thammasat University, 1980.

Published Works

Adas, Michael, *The Burma Delta: Economic Development and Social Change on an Asian Rice Frontier, 1852–1941*, Madison, Wisconsin: University of Wisconsin Press, 1974.

Allen, G. C., *A Short Economic History of Modern Japan*, 3rd ed., London: George Allen & Unwin, 1972.

Anderson, Benedict R. O'G., 'Studies of the Thai State: The State of Thai Studies', in Eliezer B. Ayal (ed.), *The Study of Thailand: Analyses of Knowledge, Approaches, and Prospects in Anthropology, Art History, Economics, History, and Political Science*, Athens, Ohio: Ohio University, Center for International Studies, Southeast Asia Series no. 54, 1978, pp. 193–247.

Ansari, Nasim, *Economics of Irrigation Rates: A Study in Punjab and Uttar Pradesh*, London: Asia Publishing House, 1968.
Ayal, Eliezer B., 'Value Systems and Economic Development in Japan and Thailand', in Robert O. Tilman (ed.), *Man, State, and Society in Contemporary Southeast Asia*, London: Pall Mall Press, 1969, pp. 535–49.
Batson, Benjamin A., [Review Article], *M. C. Sithiporn Kridakara Memorial Volume, Journal of the Siam Society*, vol. 61, pt. 1 (January 1973), pp. 302–9.
Bell, Peter F., *The Historical Determinants of Underdevelopment in Thailand*, New Haven: Yale University, Economic Growth Center, Discussion Paper no. 84, February 1970.
Binns, B. O., *Agricultural Economy in Burma*, Rangoon: Superintendent, Government Printing and Stationery, Burma, 1948.
Bowring, Sir John, *The Kingdom and People of Siam*, London, 1857, 2 vols. Reprinted, Kuala Lumpur: Oxford University Press, 1969.
Brown, Ian, 'British Financial Advisers in Siam in the Reign of King Chulalongkorn', *Modern Asian Studies*, vol. 12, pt. 2 (April 1978), pp. 193–215.
―――, 'Siam and the Gold Standard, 1902–1908', *Journal of Southeast Asian Studies*, vol. 10, no. 2 (September 1979), pp. 381–99.
―――, 'The Siamese Administrative Élite in the Early Twentieth Century and the Historical Origins of Underdevelopment in Siam', in Jeremy H. C. S. Davidson (ed.), *Lāi Sū' Thai: Essays in Honour of E. H. S. Simmonds*, London: School of Oriental and African Studies, 1987, pp. 151–90.
Campbell, J. G. D., *Siam in the Twentieth Century: Being the Experiences and Impressions of a British Official*, London: Edward Arnold, 1902.
Carter, A. Cecil (ed.), *The Kingdom of Siam*, New York and London: G. P. Putnam's Sons, 1904.
Chatthip Nartsupha, 'The Economic Thought of Phraya Suriyanuwat', in Vichitvong na Pombhejara (ed.), *Readings in Thailand's Political Economy*, Bangkok: Bangkok Printing Enterprise, 1978, pp. 402–13.
――― and Somphop Manarangsan (eds.), *Prawatsāt sēthakit thai čhonthu'ng p.s. 2484 (The Economic History of Thailand before 1941)*, Bangkok: Thammasat University, 1984.
――― and Suthy Prasartset (eds.), *The Political Economy of Siam 1851–1910*, Bangkok: The Social Science Association of Thailand, 1981.
―――, Suthy Prasartset, and Montri Chenvidyakarn (eds.), *The Political Economy of Siam 1910–1932*, Bangkok: The Social Science Association of Thailand, 1981.
Cheng Siok-Hwa, *The Rice Industry of Burma 1852–1940*, Kuala Lumpur: University of Malaya Press, 1968.
Collis, Maurice, *Wayfoong: The Hongkong and Shanghai Banking Corporation*, London: Faber & Faber, 1965.

Cushman, J. W., 'The Khaw Group: Chinese Business in Early Twentieth-century Penang', *Journal of Southeast Asian Studies*, vol. 17, no. 1 (March 1986), pp. 58-79.

HM Customs Department, Bangkok, *The Foreign Trade and Navigation of the Port of Bangkok* [1907-8 to 1920-1].

_____, *Annual Statement of the Foreign Trade and Navigation of the Kingdom of Siam* [2466 (1923/4) and 2467 (1924/5)].

Donner, Wolf, *The Five Faces of Thailand: An Economic Geography*, London: C. Hurst & Company, 1978.

Feeny, David, 'Competing Hypotheses of Underdevelopment: A Thai Case Study', *Journal of Economic History*, vol. 39, no. 1 (March 1979), pp. 113-27.

_____, 'Paddy, Princes, and Productivity: Irrigation and Thai Agricultural Development, 1900-1940', *Explorations in Economic History*, vol. 16, no. 2 (April 1979), pp. 132-50.

_____, 'Infrastructure Linkages and Trade Performance: Thailand, 1900-1940', *Explorations in Economic History*, vol. 19, no. 1 (January 1982), pp. 1-27.

_____, *The Political Economy of Productivity: Thai Agricultural Development, 1880-1975*, Vancouver and London: University of British Columbia Press, 1982.

_____, 'Extensive versus Intensive Agricultural Development: Induced Public Investment in Southeast Asia, 1900-1940', *Journal of Economic History*, vol. 43, no. 3 (September 1983), pp. 687-704.

Fisher, Charles A., *South-East Asia: A Social, Economic and Political Geography*, 2nd ed., London: Methuen, 1966.

Furnivall, J. S., *An Introduction to the Political Economy of Burma*, Rangoon: Burma Book Club, 1931.

Gleeck, Lewis E. Jr., *American Institutions in the Philippines (1898-1941)*, Manila: Historical Conservation Society, 1976.

Gordon, Robert, 'The Economic Development of Siam', *Journal of the Society of Arts*, vol. 39 (1891), pp. 283-98.

Graham, W. A., *Siam*, 3rd ed., London: Alexander Moring, 1924, 2 vols.

Great Britain, Foreign Office, *Diplomatic and Consular Reports: Siam* [various titles], London: series from 1865.

Harvey, G. E., *British Rule in Burma, 1824-1942*, London: Faber & Faber, 1946.

Hayami, Yujiro, *A Century of Agricultural Growth in Japan: Its Relevance to Asian Development*, Tokyo: University of Tokyo Press, 1975.

Hong Lysa, *Thailand in the Nineteenth Century: Evolution of the Economy and Society*, Singapore: Institute of Southeast Asian Studies, 1984.

Hyde, Francis E., *Far Eastern Trade 1860-1914*, London: Adam & Charles Black, 1973.

'Industrial Siam: The Siam Cement Company', *The Record* (The Board of

Commercial Development), no. 9, July 1923, pp. 13–15.

Ingram, James C., *Economic Change in Thailand 1850–1970*, Stanford: Stanford University Press, 1971.

Ishikawa, Shigeru, *Essays on Technology, Employment and Institutions in Economic Development: Comparative Asian Experience*, Tokyo: Kinokuniya Company; Hitotsubashi University, The Institute of Economic Research, Economic Research Series, no. 19, 1981.

Likhit Dhiravegin, 'The Meiji Restoration (1868–1912) and the Chakkri Reformation (1868–1910): A Case for a Comparative Study', in Chira Hongladarom and Medhi Krongkaew (eds.), *Comparative Development: Japan and Thailand*, Bangkok: Thammasat University Press, 1981, pp. 71–100.

Le May, Reginald, *An Asian Arcady: The Land and Peoples of Northern Siam*, Cambridge: W. Heffer, 1926.

———, *The Economic Conditions of North-Eastern Siam*, Bangkok: Ministry of Commerce and Communications, June 1932.

Mackenzie, Compton, *Realms of Silver: One Hundred Years of Banking in the East*, London: Routledge & Kegan Paul, 1954.

Maung Shein, *Burma's Transport and Foreign Trade, 1885–1914*, Rangoon: Department of Economics, University of Rangoon, 1964.

May, Glenn Anthony, *Social Engineering in the Philippines: The Aims, Execution, and Impact of American Colonial Policy, 1900–1913*, Westport, Conn., Greenwood Press, 1980.

Ministry of Agriculture, *Prawat krasuangkasēt (History of the Ministry of Agriculture)*, Bangkok, 1957.

Ministry of Commerce and Communications, *Siam: Nature and Industry*, Bangkok, 1930.

Ministry of Finance, *Report of the Financial Adviser on the Budget of the Kingdom of Siam*, Bangkok: annually from 1901–2.

Moulder, Frances V., *Japan, China and the Modern World Economy: Toward a Reinterpretation of East Asian Development ca. 1600 to ca. 1918*, Cambridge: Cambridge University Press, 1977.

Murray, Martin J., *The Development of Capitalism in Colonial Indochina (1870–1940)*, Berkeley: University of California Press, 1980.

Myers, Ramon H., *The Chinese Economy: Past and Present*, Belmont, California: Wadsworth, 1980.

Pointon, A. C., *The Bombay Burmah Trading Corporation Limited 1863–1963*, London: Wallace Brothers & Company, 1964.

Reynolds, Craig J. and Hong Lysa, 'Marxism in Thai Historical Studies', *Journal of Asian Studies*, vol. 43, no. 1 (November 1983), pp. 77–104.

Robequain, Charles, *The Economic Development of French Indo-China*, London: Oxford University Press, 1944.

Royal Department of Mines, *Notes on Mining in Siam*, Bangkok, 1921.

Ruambotkhatyọ withayāniphon prawatsāt p.s. 2488–2527 (Abstracts of Theses in History, 1945–1984), cremation volume for Professor Rong Syamananda, Bangkok, August 1985.

Sanitwongse, Dr Yai Suvabhan, *The Rice of Siam* (with a preface by HRH Prince Damrong Rajanubhab), cremation volume for Suvabhan Sanitwongse, Bangkok, 1927.

Sansom, Robert L., *The Economics of Insurgency in the Mekong Delta of Vietnam*, Cambridge, Mass.: MIT Press, 1970.

Siamwalla, Ammar, *Land, Labour and Capital in Three Rice-Growing Deltas of Southeast Asia 1800–1940*, New Haven: Yale University, Economic Growth Center, Discussion Paper no. 150, July 1972.

The Siam Cement Company (published in connection with the celebration of the year BE 2500), Bangkok: The Siam Cement Company, 1957.

The Siam Cement Company: In Commemoration of the 50th Anniversary [compiled and written by Kukrit Pramoj], Bangkok: The Siam Cement Company, 1963.

The Siam Cement Company, 'Balance Sheet Presented to the Annual Ordinary General Meeting of Shareholders', 1916–17, 1922–38. (Held in the main office of the Siam Cement Company, Bāng Sū'.)

Siam Commercial Bank, *Thīralu'k wanpōet samnakngānyai thanākhānthaiphānit čhamkat 19 singhākhom 2514 (To Commemorate the Opening of a New Head Office of the Siam Commercial Bank, 19 August 1971)*, Bangkok: Siam Commercial Bank, 1971.

Sithi-Amnuai, Paul, *Finance and Banking in Thailand: A Study of the Commercial System, 1888–1963*, Bangkok: Thai Watana Panich, 1964.

Sithiporn Kridakara, *Some Aspects of Rice Farming in Siam*, Bangkok: Suksit Siam, 1970.

Skinner, G. William, *Chinese Society in Thailand: An Analytical History*, Ithaca: Cornell University Press, 1957.

Small, Leslie E., 'Historical Development of the Greater Chao Phya Water Control Project: An Economic Perspective', *Journal of the Siam Society*, vol. 61, pt. 1 (January 1973), pp. 1–24.

Smyth, H. Warington, *Five Years in Siam: From 1891 to 1896*, London: John Murray, 1898, 2 vols.

Sturmey, S. G., *British Shipping and World Competition*, London: The Athlone Press, 1962.

Sungsidh Piriyarangsan, *Thunniyom khunnāng thai p.s. 2475–2503 (Thai Bureaucratic Capitalism, 1932–1960)*, Bangkok: Institute of Social Research, Chulalongkorn University, 1983.

Surichai Wun' Gaeo, 'The Significance of Socio-Political Factors in Meiji Economic Development with Comparative Reference to the Thai Case', in Chira Hongladarom and Medhi Krongkaew (eds.), *Comparative Development: Japan and Thailand*, Bangkok: Thammasat University Press, 1981, pp. 101–29.

Suthy Prasartset and Chatthip Nartsupha, 'The Rise of Dependent Commodity Production in Siam, 1855–1910', *The Review of Thai Social Science*, 1977, pp. 144–68.

Tanabe, Shigeharu, 'Historical Geography of the Canal System in the Chao Phraya River Delta, from the Ayutthaya Period to the Fourth Reign of the Ratanakosin Dynasty', *Journal of the Siam Society*, vol. 65, pt. 2 (July 1977), pp. 23-72.

―――, 'Land Reclamation in the Chao Phraya Delta', in Yoneo Ishii (ed.), *Thailand: A Rice-Growing Society*, Honolulu: The University Press of Hawaii; Monograph of the Center for Southeast Asian Studies, Kyoto University, 1978, pp. 40-82.

Tandrup, Anders, 'Some Danish Contributions to the Administrative and Socio-economic Development of Thailand since 1875', in *Thai-Danish Relations: 30 Cycles of Friendship*, Copenhagen: The Royal Danish Ministry of Education, 1980.

Tej Bunnag, 'Khabotngiao mu'angphrāe r.s. 121' ('The 1902 Shan Rebellion at Phrāe'), *Sangkhomsāt parithat*, vol. 6, no. 2 (September-November 1968), pp. 67-80. An abbreviated version appears in, Tej Bunnag, *Khabot r.s. 121 (The 1902 Rebellions)*, Bangkok: Thai Watana Panich, 1981.

―――, *The Provincial Administration of Siam 1892-1915: The Ministry of the Interior under Prince Damrong Rajanubhab*, Kuala Lumpur: Oxford University Press, 1977.

Thamsook Numnonda, 'The Anglo-Siamese Secret Convention of 1897', *Journal of the Siam Society*, vol. 53, pt. 1 (January 1965), pp. 45-60.

Thiravet Pramuanratkarn, 'The Hongkong Bank in Thailand: A Case of a Pioneering Bank', in Frank H. H. King (ed.), *Eastern Banking: Essays in the History of the Hongkong and Shanghai Banking Corporation*, London: The Athlone Press, 1983, pp. 421-34.

Thompson, Virginia, *Thailand: The New Siam*, New York: The Macmillan Company, 1941.

Van der Heide, J. Homan, 'The Economical Development of Siam during the Last Half Century', *Journal of the Siam Society*, vol. 3, pt. 2 (1906), pp. 74-101.

―――, *General Report on Irrigation and Drainage in the Lower Menam Valley*, Bangkok: Ministry of Agriculture, 1903.

Vella, Walter F., *Chaiyo! King Vajiravudh and the Development of Thai Nationalism*, Honolulu: The University Press of Hawaii, 1978.

Ward, T. R. J., *Report on a Scheme for the Irrigation of so much of the Valley of the Menam Chao Bhraya as may be Possible for a Capital Outlay of One and Three Quarter Millions Sterling*, Bangkok: Royal Irrigation Department, 1915.

Wedel, Yuangrat (Pattanapongse), *Modern Thai Radical Thought: The Siamization of Marxism and its Theoretical Problems*, Bangkok: Thammasat University, Thai Khadi Research Institute, Research Series no. 4, July 1982.

Wickizer, V. D. and Bennett, M. K., *The Rice Economy of Monsoon Asia*, Stanford: Food Research Institute, Stanford University, 1941.

Wilson, Constance M., *Thailand: A Handbook of Historical Statistics*, Boston, Mass.: G. K. Hall & Co., 1983.

Wittfogel, Karl A., *Oriental Despotism: A Comparative Study of Total Power*, New Haven: Yale University Press, 1957.

Wongsānupraphat, Čhaophrayā, *Prawat krasuangkasētrāthikān (History of the Ministry of Agriculture)*. First published in r.s. 129 (1910–11); republished, Bangkok: Fine Arts Department, 1941.

Wong Lin Ken, *The Malayan Tin Industry to 1914: With Special Reference to the States of Perak, Selangor, Negri Sembilan and Pahang*, Tucson, Arizona: The University of Arizona Press, 1965.

Wyatt, David K., *The Politics of Reform in Thailand: Education in the Reign of King Chulalongkorn*, New Haven and London: Yale University Press, 1969.

_____, *Thailand: A Short History*, New Haven and London: Yale University Press, 1984.

Yasukichi Yasuba and Likhit Dhiravegin, 'Initial Conditions, Institutional Changes, Policy, and their Consequences: Siam and Japan, 1850–1914', in Kazushi Ohkawa and Gustav Ranis with Larry Meissner (eds.), *Japan and the Developing Countries: A Comparative Analysis*, Oxford: Basil Blackwell, 1985, pp. 19–34.

Index

ADAMSON, SIR HARVEY, 50
Agricultural Commissioner (*khāluang kasēt*), 67–70, 82–3
Agricultural Department, 63
Agriculture: advance in techniques and practices, 6, 16, 60–76 *passim*, 79, 82–3, 86, 90 n. 44, 171–6, 180; depression, 77–88 *passim*; fairs and exhibitions, 65, 70, 72–3
Agriculture School, 65–7, 69–70
Anderson, Benedict R. O'G., 57 n. 106
Anglo-Siam Corporation, 111
Anglo-Siamese Treaty (1874 and 1883), 111–12
Animal breeding station, 65
Anuruthathēwā, Mǫm, 54 n. 37; criticism of van der Heide, 20–2, 54 n. 39
Archer, W. J., 80, 92 n. 93
Ardron, G. H., 144–5
Armed Forces, expenditure on, 41, 57 n. 106, 171, 175
Asiatic Mode of Production, 51 n. 2
Ayudhya period, canal construction in, 8
Ayuthia province, 71

BĀNG SŪ': clay deposits, 152; site for Siam Cement Company, 154
Bangkok City Bank, 136
Bangkok Times, 24, 67, 78–80, 83, 87, 92 n. 94, 140, 143, 146 nn. 9 and 18, 149 n. 89
Banking (indigenous), 6, 125–45 *passim*, 146 n. 9
Banque de L'Indochine, 125–6, 128–9, 141, 146 n. 11
Battye, Noel Alfred, 57 n. 106
Binns, B. O., 48
Bombay Burmah Trading Corporation, 111–12, 116–17
Book Club, 128–35, 146 nn. 18 and 27, 147 n. 45; *see also* Siam Commercial Bank

Borneo Company, 110–12, 116, 156
Bowring Treaty (and nineteenth-century commercial treaties in general), 9, 51 n. 3, 105–7, 110, 176–7; removes fiscal autonomy, 35, 38, 175; sanctioning of mining legislation, 100–1, 104
Bucket dredge, 101–2, 121 n. 49, 176
Burīram, 159
Burma: advance in agricultural techniques and practices, 74–5, 176; irrigation works, 47–51, 59 n. 141; rice exports competing with Siam, 26, 28, 30, 34, 36–7, 43, 71–2, 79, 174; teak industry, 94, 111

CANAL DEPARTMENT, 18–27, 42, 54 n. 39, 55 n. 47, 66
Cattle trade (from the north-east), 163–4
Cement industry, 151–5 *passim*, 157, 166
Chaināt, 16, 19, 115
Čhanthaburī, Prince: cotton cultivation, 165; irrigation, 25, 29, 30–2, 38–40; Siam Commercial Bank, 141–4; Siamese Steamship Company, 156
Čhaophrayā delta (comparison with Irrawaddy and Mekong deltas), 2, 4, 47, 49, 51, 53 n. 19
Chartered Bank, 125–9, 131–3, 141, 146 n. 11
Chatthip Nartsupha, 55 n. 52, 170–4, 177, 182 n. 1
Cheek, Marion A., 112–13, 119
Chiangmai, 30, 57 n. 107, 111–15
China, silk industry, 162–3
Chinese labour, 2–3, 9; gang riots in tin industry, 103–4, 122 n. 66; employed in tin industry, 96–7
Chinese merchant community: as dependent bourgeoisie, 171–4, 177, 182 n. 5; investment in indigenous banks, 128–30, 137, 177; rice millers and traders, 126, 136–7, 140, 142;

INDEX 193

Siamese Steamship Company, 155, 168 n. 34, 177; tin mining entrepreneurs, 96–9, 102, 116, 121 n. 49

Chino-Siam Bank, 136–44, 150 n. 118

Čhitčhamnongwānit, Luang, 137, 139, 141

Chǫng Khāe, limestone deposits, 152–3

Chulalongkorn, King: advance in agricultural techniques and practices, 68–9, 72; cement company, 152; central revenue office (1870s), 171; irrigation, 18–19, 42, 174; Nāi Čharoen, 61–3, 89 n. 15; owns land in Rangsit, 46; Prince Mahit and the Siam Commercial Bank, 128, 131–4, 146 n. 27, 147 nn. 45 and 48; Prince Phenphat, 65; Rangsit depression, 82–5, 88; railway construction, 15; sericulture initiative, 161; tin industry, 95–8, 100, 104, 108–9; van der Heide, 13–14, 24

Chumphǫn, 106

Civil Service College, 66–7

Cochin-China: advance in agricultural techniques and practices, 49, 74–5; irrigation works, 47–51, 59 nn. 128 and 139; rice exports competing with Siam, 26, 28, 30, 43, 71–2, 79, 174

Cotton cultivation, Government initiative in, 151, 157, 164–6

DAMNOEN SADUAK CANAL, 10, 18, 21, 54 n. 39

Damrong, Prince, 62, 66, 90 n. 44, 99, 148 n. 57, 158; depression in Rangsit, 77, 82, 85; irrigation, 18; tin industry, 97, 100, 103–4, 106–8, 117

Danish East Asiatic Company, 111, 152–3

Danske Landmannsbank, 130, 146 n. 9

De la Mahotière, L. R., dispute with van der Heide, 23–4, 54 n. 41

De Müller, Walter, 97, 99, 103

De Richelieu, Commodore, 104

Denmark, exports cement to Siam, 151–2

Department of Mines, 95, 97, 99, 100, 103, 106–7, 109, 116–17; European officials in conflict with Prince Damrong, 106–7, 117

Deutsch-Asiatische Bank, 130–1, 133, 135–8

Devawongse, Prince, 62, 113, 133, 146 n. 9; irrigation, 17–18

EASTERN SMELTING COMPANY, 102

European powers, threat to Siamese sovereignty, 1–2, 7, 41, 57 n. 106, 104–5, 110, 112, 128, 175, 179

Exchange (baht), 79, 83, 85, 88; adoption of gold-exchange standard (1902), 126–8, 145 n. 7; exchange business (banks), 126–33, 136

Exchange reserves (baht), 40–2, 57 nn. 101 and 108, 175

Exhibition of Agriculture and Commerce (1910 and 1911), 72–3

Experimental farm, 61–3, 65–7, 70–3, 82, 91 n. 52; Bangbert, 90 n. 44

FENNY, DAVID, 34–6, 38, 40, 44–6, 58 nn. 114 and 116, 71–2

Financial Adviser, British, influence of, 34, 36–7, 53 n. 25, 54 n. 36, 56 n. 87, 127, 129

Forest companies (European), 94, 110–12, 114–19, 124 n. 139, 176

Foresters (Burmese and Shan), 110–11, 118–19

Forestry Department, 114–19, 124 nn. 124 and 129; European officials in, 117

Forestry legislation and leases (post-1896), 114–19

French East Asiatic Company, 111

French Indo-China: cement industry, 151; silk industry, 162–3

GAMBLING, EFFECT ON PEASANTRY, 79, 81, 84, 86–7

Gollo, E. G., 152

Graham, W. A., 90 n. 41, 93 n. 113, 161

Grassi, Joachim, 11

Greville, George, 146 n. 9

Grut, Captain W. L., 153–4, 167 n. 15

HANDICRAFT PRODUCTION, 73, 87, 171, 179–80

Hannen, Sir Nicholas, 113

Hongkong and Shanghai Bank, 125–9, 131–3, 139, 141, 145, 146 n. 11; van der Heide account, 23, 25

IMPERIAL FOREST COLLEGE (DEHRA DUN), 116
Ingram, James C., 33–4, 37, 40
Irrawaddy Delta, 2, 4, 47–9, 51
Irrigation and drainage, 6, 8–59 *passim*, 60, 71–2, 76–9, 81, 83, 87, 92 n. 94, 171–5
Irrigation Department, 13–14, 18, 28–9, 31–2, 37, 44–5, 58 n. 113, 79
Irrigation charges, 17, 31–3, 36–9, 78; in British India, 56 n. 88

JAPAN: advance in agricultural techniques and practices, 64–5, 75–6, 91 nn. 69 and 70, 176; nineteenth-century economic transformation (comparison with Siam), 178–80, 183 n. 11; silk industry, 162–3
Johnston, David, 11, 36, 40, 58 n. 115, 92 n. 74, 93 n. 116
Joo Seng Heng, 136–44, 148 n. 72, 149 n. 89, 150 nn. 116, 118, 119 and 122

KHŌRĀT, 24; sericulture, 157–60
Kilian, Felix, 130, 134
Koch (of Deutsch-Asiatische Bank), 137
Koh Tiew Lim, 108–9
Kwong Yik Bank (Singapore), 139

LAMPĀNG, 30, 111; Shan attack, 15
Lamphūn, 111
Land bank (agricultural bank), 79–81, 84, 87, 92 n. 94
Land taxation, 38–9, 78–80, 82–5, 88; limited by treaty, 35, 38
Lao labourers, 81–4
Le Count, W. K., 145
Le May, Reginald, 160
Leonowens, Louis T.: cement interests, 151; teak interests, 111
Lloyd, W. F., 124 n. 129
Loans (foreign), 35, 40, 175; proposed loan for irrigation, 27, 29–32, 39–40, 44; 1905, 15, 39; 1907, 40; 1909, 106; 1922, 40; 1924, 40
Lopburī, 15, 41

MAHĀSAWAT CANAL, 9, 10
Mahit, Prince, 15, 62, 100, 126; Book Club/Siam Commercial Bank, 127–35, 146 nn. 18 and 27, 147 nn. 45, 47 and 48; irrigation, 18
Malay States: imports Siamese cement, 154; tin industry, 94–101, 116, 120 n. 16, 121 n. 49
Mekong Delta, 2, 4, 47–9, 51
Miles, Captain Edward T., 95–6, 104, 121 n. 45
Mines Commissioner, 99
Mining companies (European), 102–10, 116, 123 n. 90, 176
Mining legislation, 98–101, 104, 109, 123 n. 90
Ministry of Agriculture, 23, 25–6, 28, 61–3, 65–8, 70, 72, 83, 85, 90 n. 41, 99, 107, 158, 164
Ministry of Communications, 26, 28
Ministry of Finance, 23, 25, 82–3, 85, 88, 142, 145, 168 n. 34
Ministry of the Interior, 66, 77–8, 82–3, 85, 87–8, 90 n. 41, 103, 107, 114–15
Ministry of Local Government, 82–3, 85, 88
Ministry of Public Works, 85, 152
Mongkut, King, canal construction during reign of, 9, 10
Murray, Martin J., 49

NĀI ČHAROEN, 61–3, 65, 70, 89 n. 15
Nakhǫn Chaisī, 9, 10, 63
Nakhǫn Pathom, 9, 65, 91 n. 52
Nakhǫn Sawan, 15
Nakhǫn Sīthamarāt, 99, 106
Nān, 15
Narāthip Praphanphong, Prince, 46, 112–13
Narit, Prince, 18

PAGET, RALPH, 131–4, 147 n. 47
Pāknamphō, 115
Parkes, Sir Harry, Supplementary Agreement of 1856, 38
Pāsak, 19
Pāsak scheme, 28, 31–3, 35, 42, 44, 58 nn. 113 and 114, 173
Pattānī, 106
Peasantry: 'indolence and irresponsibility of', 81, 84, 86–7,

INDEX

93 n. 113, 160–1; response to sericulture initiative, 159–61, 163–4, 166
Phāsī Čharoen Canal, 9, 18, 21, 51 n. 4, 54 n. 39
Phāsī Sombatbǭribūn, Phra (Pho Jim), 10, 51 n. 4
Phenphatanaphong, Prince, 63–70, 89 n. 20, 91 n. 52, 158, 160–1, 163, 180
Phichit Prīchākǫn, Prince, 46
Philippines, advance in agricultural techniques and practices, 74, 176
Phitsanulōk, 15, 91 n. 52; cotton cultivation in, 164–5
Phrāe, 119; Shan attack (1902), 14–15, 29, 41
Phūket, 95–9, 101, 103–4, 106–8, 122 n. 66, 123 n. 87, 178
Phūket harbour, 95–6, 121 n. 45
Prajadhipok, King, 145
Prakǭp kasikam (Agriculture), 72–3
Prawēt Canal, 18
Privy Purse Department: Siam Cement Company, 153–4, 167 n. 14; Siam Commercial Bank, 128, 130, 137, 141–5; Siamese Steamship Company, 168 n. 34

RAILWAY CONSTRUCTION, 3–4, 6, 15, 17–18, 24, 29–31, 33–7, 41–2, 56 n. 68, 57 n. 107, 106, 108, 151, 175
Rama III, canal construction during the reign of, 8–9
Rāmāthibǭdī, King, canal construction during the reign of, 8
Rangsit district, 12–14, 16, 19, 32, 35, 44–5, 58 nn. 113, 114, 115 and 116, 70–2, 78–9, 81–2, 84; Rangsit in depression, 77–88, 92 n. 74, 93 n. 116; Rangsit experimental farm, 71–3, 91 n. 52; Rangsit landowners, 35, 44–6, 54 n. 37, 58 n. 116, 70–1, 77–80, 82–4, 87–8, 173–4
Ranǭng, 103, 108
Rātburī, 65, 91 n. 52
Rātburīdirēkrit, Prince, 26–8, 30–2, 37–40, 43, 48, 71, 165, 174
Regulation of Canal Excavation (1877), 10–11, 37
Rice: advance in techniques and practices of cultivation, 16, 60–76 *passim*, 79, 82–3, 86, 90 n. 44, 171, 173–6; rice economy in depression, 77–88 *passim*, 137, 181
Rivett-Carnac, Charles: abandonment of silver standard, 126; irrigation, 13–15, 53 n. 25
Robequain, Charles, 48–50
Rolin-Jacquemyns, 14
Royal Pages School, 66

SĀEN SĀEP CANAL, 18
Sāen Sāep district, 82
Sai Sanitwongse, Phra-ongčhao, 11, 46, 54 n. 37, 70, 83
Saigon, visit of Nāi Čharoen, 62–3
Samrōng Canal, 18
Samut Sākhǭn, 10
Sansom, Robert L., 49
Saraburī district, 154
Savings bank (rural), 81, 83, 87
Sawatdīwiangchai, Luang, 153
Schultz, Oscar, 153
Schwarze (of Deutsch-Asiatische Bank), 137–8, 148 n. 57, 150 n. 122
Scott, H. G., 97, 107
Secret Convention (Anglo-Siamese, 1897), 104–5
Sericulture: government initiative, 151, 157–64 *passim*, 166; Japanese experts, 63, 65–6, 158–60, 162; 'overbearing manner of officials', 161; Siamese students of, 66, 159–60
Sericulture Department, 158–63
Sericulture School, 66, 159
Siam Cement Company, 151, 153–5, 166, 167 n. 30, 173, 177–8
Siam Commercial Bank, 125, 134–45, 148 n. 57, 149 n. 89, 150 nn. 122 and 129, 151, 177–8; *see also* Book Club
Siam Electric Company, 153
Siam Land, Canals, and Irrigation Company, 11–13, 26–7, 32, 45–6, 54 n. 37, 77–9
Siam Rice Milling Company, 137, 139, 141
Siamese administrative élite, 7, 42, 70, 80, 125, 143, 151, 157, 170, 179–81, 182 n. 5; economic interests of, 5–6, 34–5, 44–6, 54 n. 37, 58 nn. 116 and 120, 71–2, 155, 170–4, 177–8

Siamese Steamship Company, 151, 155–7, 166, 168 n. 34, 177–8
Sīsuriyawong, Somdet Čhao Phrayā (Chuang Bunnag), 10
Sithiporn Kridakara, Mǫm čhao, 90 n. 44, 180
Slade, H. A., 114–17
Small, Leslie E., 36–7, 40, 55 n. 47
Smidth, F. L., 154
Smyth, H. Warington, 95, 97–100, 103, 122 n. 70
Sōphonphetcharat, Luang (Kim Seng Lee), 136
Srīsunthǫnwōhān, Phrayā, 83
Straits Trading Company, 102
Suphan scheme, 28, 32–3, 35, 44
Surasakmontrī, Phrayā, 100
Suriyānuwat, Phrayā, 183 n. 14
Survey Department, 66
Suthy Prasartset, 55 n. 52, 182 n. 1
Suvabhan Sanitwongse, 46, 70, 72, 78–9, 90 n. 44, 155

TAX FARMS AND FARMERS, 38, 41, 137, 172; investment by tax farmers in indigenous banks, 128–30, 136
Taylor, C. W., 145
Teak industry, 3–4, 6, 94–5, 110–19 *passim*, 124 nn. 121 and 139, 176–7
Textile industry, proposals for, 165, 173
Thanyaburī district, 70, 72, 80–1, 87, 92 n. 74
Thēsāphibān system of provincial administration, 99, 103
Thēwētwongwiwat, Čhaophrayā: irrigation, 13–14, 17–19, 25, 36–7, 42, 174; Nāi Čharoen, 61–2, 70; Prince Phenphat, 64–5; tin industry, 98; van der Heide, 17–18, 23–4
Thipkōsā, Phrayā, 96
Tin industry, 3, 6, 94–110 *passim*, 116, 119, 119 n. 2, 123 n. 90, 176–8
Tin smelting, 102–3, 108–10, 122 n. 87, 173, 176, 178; European smelting companies, 102–3, 108–10, 176, 178
Tongkah Harbour Tin Dredging Company, 95–6, 101, 104, 121 nn. 45 and 49

Toyama Kametaro, 63, 65, 70, 158, 162–3

UNITED STATES–SIAM TREATY (1856), 113

VAJIRAVUDH, KING, 26, 28–31, 39, 107–9, 155; Siam Cement Company, 153; Siam Commercial Bank, 138, 140–2
Van der Heide, J. Homan: arrives in Siam, 14; criticism of his methods, 20–4, 27, 36, 42–3; leaves Siam, 25, 43; views on political issues, 24; writings, 55 n. 52; 1903 proposals, 15–18, 29, 33, 36–7, 40–5, 52 n. 19, 53 nn. 20 and 25, 57 n. 102; 1906 and 1908 proposals, 19–20, 42, 44, 53 n. 29, 55 n. 47, 57 n. 109

WARD, SIR THOMAS, 28–33, 37–40, 42–5
Water conditions, for rice cultivation, 16, 21, 23, 26, 33–4, 47, 52 n. 19, 54 n. 39, 55 n. 47, 79–82, 84, 87
Westengard, Jan, 80, 131, 133–4, 138, 147 n. 47
Willeke (of Deutsch-Asiatische Bank), 144
Williamson, W. J. F., 148 n. 57; irrigation, 20, 25, 32, 34, 36–8, 58 n. 114
Wilson, R. C. R., 28, 31
Wittfogel, Karl A., 51 n. 2
Wongsānupraphat, Čhaophrayā, 30, 46, 88, 180; advance in agricultural techniques and practices, 67–70, 73; depression in Rangsit, 80–8, 93 nn. 108 and 113; tin industry, 107–9
World Grain Exhibition and Conference (Regina, 1933), 73

YOMARĀT, ČHAOPHRAYĀ: Siam Cement Company, 152–3, 155, 167 n. 14, 177; Siam Commercial Bank, 139–41

EAST ASIAN HISTORICAL MONOGRAPHS

British Mandarins and Chinese Reformers:
The British Administration of Weihaiwei (1898-1930) and the Territory's Return to Chinese Rule
Pamela Atwell

The Chinese Communists' Road to Power:
The Anti-Japanese National United Front, 1935-1945
Shum Kui-Kwong

Chinese Politics in Malaysia:
A History of the Malaysian Chinese Association
Heng Pek Koon

Democracy Shelved: Great Britain, China, and Attempts at Constitutional Reform in Hong Kong, 1945-1952
Steve Yui-Sang Tsang

The Élite and the Economy in Siam c. 1890-1920
Ian Brown

The Emergence of the Modern Malay Administrative Élite
Khasnor Johan

The Federal Factor in the Government and Politics of Peninsular Malaysia
B. H. Shafruddin

Hong Kong under Imperial Rule, 1912-1941
Norman Miners

Malay Society in the Late Nineteenth Century:
The Beginnings of Change
J. M. Gullick

The Origins of an Heroic Image:
Sun Yatsen in London, 1896-1897
J. Y. Wong

The Peasant Robbers of Kedah 1900-1929:
Historical and Folk Perceptions
Cheah Boon Kheng

The Structure of Chinese Rural Society:
Lineage and Village in the Eastern New Territories, Hong Kong
David Faure

Thai-Malay Relations: Traditional Intra-regional Relations from the Seventeenth to the Early Twentieth Centuries
Kobkua Suwannathat-Pian

A Wilderness of Marshes:
The Origins of Public Health in Shanghai, 1843-1893
Kerrie L. MacPherson